LAST AMERICAN HEROES
TODAY'S FIREFIGHTERS

**CHARLES W. SASSER
and
MICHAEL W. SASSER**

POCKET BOOKS
New York London Toronto Sydney Tokyo Singapore

The sale of this book without its cover is unauthorized. If you purchased this book without a cover, you should be aware that it was reported to the publisher as "unsold and destroyed." Neither the author nor the publisher has received payment for the sale of this "stripped book."

An *Original* Publication of POCKET BOOKS

POCKET BOOKS, a division of Simon & Schuster Inc
1230 Avenue of the Americas, New York, NY 10020

Copyright © 1994 by Charles W. Sasser and Michael W. Sasser

All rights reserved, including the right to reproduce
this book or portions thereof in any form whatsoever.
For information address Pocket Books, 1230 Avenue
of the Americas, New York, NY 10020

ISBN: 978-1-4767-8448-9

First Pocket Books printing August 1994

10 9 8 7 6 5 4 3 2 1

POCKET and colophon are registered trademarks of
Simon & Schuster Inc.

Cover photo by Steven Spak

GIVE US THE RIGHT EQUIPMENT AND WE'LL GO RIGHT INTO HELL AND KICK SATAN'S ASS . . .

Fire roared out of the deli as the heavy bolt of water ripped into its churning guts. A solid wall of bright flames snapped at the firefighters and hurled smoke as though to choke those it could not reach.

"We're going in! Get ready," Barrett yelled above the crackling flames. His line would launch the main attack, while the second line on Engine Two and the two lines from Engine Four would cover exposures. . . .

"Barrett, got your will made out?" Bob Plane asked lightly. "Hope you got your insurance made out to me."

They joked about it, about death in the line of duty, but every firefighter knew the big stats on their profession—four out of every ten of them injured each year, nearly two hundred KIA. Killed in Action—another military term. Each time they entered a fire they challenged the stats . . . and tempted fate.

Barrett thought he heard the dry, hollow laughter of Fate as, pipe charged, he and his crew advanced behind their stream of water toward the flames. . . .

PRAISE FOR CHARLES W. SASSER'S NOVEL
THE 100TH KILL

"Few writers know men or war as well as Chuck Sasser. This is a marvelous rendering of both."
—Jim Morris, author of *War Story*

"Grisly and gripping. . . . Sasser has caught the essence of sniper warfare in Vietnam."
—Craig Roberts, author of *Police Sniper*

Other Books by Charles W. Sasser

Shoot to Kill
Always a Warrior
Homicide!
The 100th Kill
One Shot—One Kill
 (with Craig Roberts)
The Walking Dead
 (with Craig Roberts)

Published by POCKET BOOKS

For orders other than by individual consumers, Pocket Books grants a discount on the purchase of 10 or more copies of single titles for special markets or premium use. For further details, please write to the Vice-President of Special Markets, Pocket Books, 1230 Avenue of the Americas, New York, NY 10020.

For information on how individual consumers can place orders, please write to Mail Order Department, Paramount Publishing, 200 Old Tappan Road, Old Tappan, NJ 07675.

*This book is dedicated to my sons and to James R. Reilly,
deceased firefighter, friend, and former father-in-law*

Charles W. Sasser

And to my adopted son, Quincy

Michael Sasser

Acknowledgments

The authors wish to thank the people who made this book possible. Their help in the monumental task of researching and writing this book made a difficult project an enjoyable experience.

Our thanks to:

Mayor Seymour Gelber and the City Council of the City of Miami Beach, Florida;

Fire Chief Brainaird Dorris of the Miami Beach Fire Department;

Gene Spear, retired MBFD who provided technical advice and encouragement;

Bill and Bea Whipple, whose encouragement in this project helped get it started.

And, of course, a special thanks to all the brave firefighters of the Miami Beach Fire Department, the stories of many of whom appear in this chronicle.

Authors' Note

The information in this book comes from the accounts of the men and women whose stories appear on these pages. The chapters are based on taped personal interviews and upon actual observances and some participation by the authors. Actual names are used except in the instances of victims of disaster whose public identification would serve no useful purpose.

In certain instances dialogue has by necessity been re-created to match the situation, action, and personalities of those people involved, while maintaining factual content as it was presented. Also, here and there, events and persons involved in those events may not correspond precisely in the memories of all involved. Time has passed since the occurrence of some of the action related in this book. Time has a tendency to erase memory in some areas and to enhance it selectively in others.

The material has also been filtered through two authors, who must apologize to anyone omitted or neglected in the preparation of this book. Our objective was

Authors' Note

to present the selfless and heroic service of men and women who combat fires in urban America. To that end, we are confident we have neglected no one. The content remains as accurate to the spirit and reality of the American fire and rescue service as the authors can make it.

We have taken the literary license to condense the action in this narrative to fit into the Christmas holiday time frame, the busiest time of the year for most city firefighters. Story and content are arranged in such a way as to satisfy both those interested in the technical and mechanical aspects of fires and firefighters, and those who are thrilled by the personal adventures of the brave men and women who risk their lives in the never-ending war against disaster.

Finally, this book centers on a single group of firefighters at a single firehouse—Station 2—of the Miami Beach, Florida, Fire Department. Unfortunately, it is impossible to include the stories of every firefighter in the department, or even all those of Station 2, but the authors trust that the courage, professionalism, and selflessness of firefighters selected to appear in this book are an accurate reflection of the Miami Beach Fire Department and of firefighters throughout America.

"Just tell me one boy, just one who at seven didn't aspire to be a hero or a fireman."

—Miguel Delibes, *The Stuff of Heroes*

LAST AMERICAN
HEROES

1

THE FONTANA HOTEL FIRE STARTED AT ABOUT FOUR A.M., just before dawn, which was always the worst possible time for disaster in the older hotels on Miami Beach, Florida. Retirees on a budget, attracted to lower rates, slept soundly at that hour. The hotel cut its staff to a minimum after midnight. The drowsy hotel clerk held his chin with the palms of his hands to keep it from dropping onto his chest; the bellhop napped in a chair in the lobby.

The Fontana dated back to before Jackie Gleason, back to when that seven-mile-long spit of sand and palms in South Florida's Biscayne Bay known as Miami Beach was synonymous with glamour, stars, and wealth. The northern island along the bright strip of Collins Avenue had glittered with hotels and resorts, with youth and money and excitement. A call girl could go for five thousand dollars a night. When Jackie Gleason was King, he tagged Miami Beach the "Dream Capital of the World."

But then came the riots and urban unrest, and the

beginning of urban blight in the late 1960s. Jackie Gleason died, taking with him the Dream. The glitter that was Miami Beach in its heyday remained, but it was fading. Young developers poured millions into high-rise condos and into razing South Beach, that graveyard of low rents and fixed-income retirees, and renovating the old art deco hotels. In spite of all efforts, however, Miami Beach remained an aging though haughty dowager with a facelift and a thick coat of makeup.

Behind the facelift and the makeup, the island appeared about to sink from the weight of high-rise condos and fifty-year-old hotels where a "towering inferno" was always possible, by gray enclaves of lonely and abandoned retirees existing on the fringes of cluttered slums occupied by recent Caribbean refugees, by a population of some 120,000 people that increased by as much as twenty percent during peak season.

It was peak season when the Fontana Hotel fire started. Constructed of Dade County pine that had had all the moisture sucked from it by years of tropical sun, the hotel rose three stories tall and was narrow across the front, squeezed between the Prince Hamlet Hotel on the south and a ramshackle rooming house on the north. While the residents slept and the employees dozed, fire by some source sneaked into the dry recesses of the old building. Flames traveled silently like cat burglars through the walls and in the cocklofts. They slipped along air vents and plumbing tunnels and heating ducts.

While the residents slept, the intruder built up rapidly and clandestinely in the hidden recesses of the hotel. And then, as from ambush, it exploded into view. Within minutes of emerging, even while the alarm was sounded, fire involved all three floors and the lobby, cutting off and trapping enclaves of residents. Pine wood cracked like strings of firecrackers as it burned. Flames licked over the high rooftop, pumping a column of smoke, while fierce fire glowed behind the rows of windows. Some of the windows were already popping from the built-up pressure.

Station 2's engine company had had a pushout on an oven fire earlier, a garbage run. Firefighters had just stripped and crawled into their bunks when the alarm threw them onto the floor again. *"Fontana Hotel!"* the radio dispatch intoned rapidly through the in-station intercom. *"People possibly trapped inside."*

Then came a long list of assignments: *"Engine Two, Engine Three, Ladder One, Rescue Two . . ."*

"It's a Pearl Harbor!" David Mogen shouted, hopping on one foot to get his turnouts on, and then taking to the stairs. "The whole city must be afire."

Division Chief Bill Miller, a crusty veteran, sniffed the smoke from blocks away as he sirened his command vehicle to the scene. He would take charge of one of Miami Beach's most deadly fires. The entire building was already at risk. A single glance told him that. Flames bent and lapped out windows, around eves, and tongued through the gabled roof. Tremendous heat blasting out of the lobby, as from the open door of a giant furnace, ignited some cars and a motorcycle parked in front at the curb. The paint peeled and then the oil and gas in the engines caught, adding to the commander's problems.

Fire threatened the entire block. Miller faced another San Francisco fire if flames skipped to the rooming house and the Prince Hamlet and kept going along the street, eating from building to building in a conflagration that might literally consume half of Collins Avenue. He would have considered a surround and drown, except that there were people trapped inside. That meant he had to cover exposures while at the same time mounting an interior assault to grab survivors.

A fire command officer could be compared to a general commanding troops in battle. He had much the same duties and responsibilities. The lives of his own troops, and of innocent people as well, depended upon the soundness and timeliness of his judgment. It was he who planned the strategy and tactics, who softened up the enemy here with heavy artillery, withdrew there to a defensive position, or probed or attacked or scouted into enemy territory, and then finally, perhaps, launched an all-out offensive to drive the enemy into submission. Fighting a fire as big as the one at the Fontana demanded risk and timing and good judgment as well as courage and boldness.

The fire officer in charge of a blaze like the Fontana followed a basic eight-step procedure pattern in his planning. Many if not all the steps might be inaugurated concurrently, not sequentially, since speed was the most important weapon in fighting fire.

First, he had to size up the situation, which was actually an ongoing process. Preplanning inspections of the larger buildings and areas within a district provided fire officers with good intelligence on such things as available water supply, accessibility to the buildings, the dimensions and construction of buildings, access roads, occupancies and life hazards, and the possibilities of communication or spread of fire. To this preplanned intelligence he added current data—location of the fire and its extent; time of day or night, which provided traffic patterns, both pedestrian and vehicular; occupancy hazards, such as were the residents most likely to be sleeping or out working, or whatever; weather, such as wind conditions, rain, high or low humidity, all of which affected the intensity and speed of flames.

As for the Fontana, it was being taken fast. Air vents, crawl spaces, heating and air-conditioning ducts served as chimneys, as travel routes for the self-propagating flames. Well-ventilated, fueled, the fire was traveling, glowing in all the hotel windows. Flames tongued through the south walls, licked at the Prince Hamlet next door.

Even as apparatus—fire equipment—arrived, elderly people wrapped in bedding or dressed in a mismatch of whatever was at hand silently gathered across the street from the hotel, where, stunned, they blinked in disbelief. Evacuees resembling war refugees trickled out the back of the hotel. Some of them ran and screamed and called the names of those from whom they had become separated.

"Jesus! Jesus! Lord Jesus, help them!" rose a plaintive cry that hung in the dawn air like smoke. "There are more people inside. It's killing them."

A commander's second step was to make sure he had all the help he needed, not just in apparatus and combat troops, but also in auxiliary help—police, ambulances, Rescue. Hospitals might have to be alerted, the water department and the electric company notified.

The Fontana was a three-alarmer. The sounding of the third alarm stripped all four Miami Beach firehouses of their apparatus. The City of Miami across the causeway automatically dispatched fire companies to man the abandoned Beach fire stations. That left Chief Miller free to battle a big blaze with his combined task force without having to worry about a second mission if one came up.

Pumpers, trucks, Rescue, and ambulances descended upon the Fontana, sirens blowing. Police cordoned off the blaze, isolating it, while Fire Rescue established triage and aid stations and evacuation points. Chief Miller quickly decided to make it a two-front attack, from the front and the rear, with left and right exposure coverages to contain the blaze, and an aerial assault. He plotted tasks and assigned points of attack.

Within ten minutes after their arrival, pumper crews had laid two main lines in front and another in back, charging them with water. Hard master streams penetrated the smoke, hissed and steamed. Driven back from the hotel by the tremendous heat, the crews kept pumping nonetheless. Truckies dogged ladders to windows, but the flames forced them back before they could launch rescue attempts. Water poured from the aerial ladder as it rose into the air. Hydraulic jacks steadied its base as it lifted men and their water cannon toward the flaming hotel roof.

The fire created its own wind, sucking in surrounding oxygen with the velocity of a small gale. An empty Pepsi can rattled down the street and was sucked into the fire.

"My wife! My wife's in there. Help her!" rose a plaintive cry.

An elderly man in faded blue pajamas rushed toward the hotel. A policeman tackled him. People driven by terror, desperation, and concern sometimes turn into zombies and mindlessly charge back into a fire. They have to be caught and shaken to their senses, reassured that firefighters' first priority is to save lives.

Rescuing survivors was a commander's third step, but always at the top of his list. If saving a single life meant permitting an entire city block to burn, if those were the only choices, then the block would just have to go. Lives always came first.

The Fontana demanded an inside rescue. Inside

rescues, especially in a fire like this one, always proved difficult and hazardous. There was the fire itself, of course, with the threat of backdrafts and flashovers and falling roofs and ceilings. Finding victims panicked or unconscious meant rescuers had to literally search every room, every closet. Frightened children or old people often crawl underneath beds or tables or hide at the backs of closets.

While Engine 2 attacked the front of the hotel with its main line, attempting to open a rescue path, Engine 1 at the back of the building grabbed several more survivors, whom firefighters were escorting to arriving ambulances. Chief Miller saw a firefighter gently drape a blanket over a sobbing woman whose clothing had been singed off her frail body.

It was a hell of a risk—sending men into that inferno. But Miller had no other choice, and he knew his troops would have it no other way. He felt a great surge of pride in them and their courage. Like the firefighters always said, "Into the breach, goddamnit. That's why we get paid the big bucks."

The rescue had to come immediately or not at all—as soon as the main line opened a path inside. Truckies and some of the pumper men donned SCBAs —self-contained breathing apparatus—and tanks, preparing to go in to risk their own lives for the lives of those trapped inside.

Step four in the planning process was to confine the fire by covering exposures. The Miami Beach Fire Department operated aggressively, on the presumption that the best defense was a good offense. Heavy main streams soaked adjacent buildings, while other firefighters patrolled the area with booster lines to control flying embers. Once troops gained access to the building, within twenty minutes after their arrival, they fought the fire room by room, opening up ceilings and plumbing to check for fire chases, denying flames to areas not already involved.

In step five, ventilation and forcible entry, Chief Miller realized that the most he could expect was to open vents in the building to draw heat, smoke, fire, and gases away from the trapped survivors inside. Truckies chopped holes in the roof, creating chimneylike drafts to draw the smoke up and out so rescuers below could penetrate more quickly and safely. Ventilation also prevented flashovers, which could, with the sudden introduction of oxygen, transform a smoldering blaze into a fire bomb. Flashover fire, backdrafts, were every firefighter's worst nightmare.

Extinguishing the fire was the sixth step. After troops confined the blaze and covered exposures to prevent its spread, not only from building to building but also from room to room, even from area to area, assault crews attacked with every resource they had to leave the fire sogged and lifeless. There could be no compromise with this enemy; victory had to be total.

At this point in the Fontana fire, with people trapped and the fire roaring out of control, Chief Miller could not even think of the last two steps of the planning process—step seven, overhauling; and step eight, salvage.

After the fire was almost out, *when* it was out, Chief Miller would be able to release some units to prepare for the next alarm while the remaining firefighters conducted overhaul. They extinguished remaining small pockets of fire, scattered debris to make sure no embers remained, removed unsalvageable materials, and in general made certain that the fire was out and could not restart.

Every operation of a well-trained fire department starts with step eight, salvage, as its objective—both of lives and of property. Salvage is a continuing process, which means doing the least damage possible to a structure and its contents. When practical, firefighters drape furnishings with salvage covers to prevent water and smoke damage, even as the battle rages. They use minimum water, ventilate by windows and doors whenever they can, rather than hacking through walls and roofs, and in every way attempt to avoid destroying in the process of saving.

"What you have to understand," said Jim Barrett, "is that you go into a family's house, you're an intruder into their lives just as the fire is. The fire is going to cause enough trauma to the people; you don't want to add to it."

Those were the eight steps in combating fire. Chief Miller remained at step three. Rushing about assigning tasks and checking progress, he paused to blink at the thermofire. It reflected red against his ruddy face. Apparatus had been on the scene less than ten minutes; already lines had been laid and ladders dogged.

From somewhere deep and lost in the fiery guts of the inferno came a faint cry for help.

"Lay pipe into the lobby," Miller shouted. "We're going in there."

2

ACCORDING TO GREEK MYTHOLOGY, THE GOD PROMETHEus created the first man and taught him to walk upright. Zeus, the ruling Greek god, held less pride in man and decreed that man must eat his food raw. Prometheus took pity upon the frail and vulnerable creature he had created and stole the divine fire from the heavens to improve conditions on earth.

Mythology notwithstanding, it was probably lightning and not Prometheus that brought fire to earth. Men must have been terrified at the unchecked forest fires that first swept across the earth, consuming everything in their way. Mankind's first deliberate use of fire probably occurred around 500,000 B.C. Since then, fire as friend has advanced with civilization. Today's harnessing of atomic energy is only one of the latest uses of fire.

Fire as foe has also scourged man's advance through civilization. Nearly twenty thousand lives are lost yearly in the United States alone. Fire accounts for *billions* of dollars in damage. While fire is friend, man discovered, it is also enemy. Men sought it for what it could do for them; they also sought ways to harness it and prevent its destroying what they had created.

The first record of regulating building safety as a

means of fire protection was written into law during the reign of Hammurabi of Babylonia in about 2000 B.C. Hammurabi declared that any builder whose structure collapsed in fire or otherwise because of faulty construction should die. Centuries later, in 24 B.C., the Roman Emperor Augustus took the first step in providing fire protection by creating a *vigile,* slaves who patrolled the city, checking for fires.

The custom of having a night fire watch spread from city to city, nation to nation. In 1000 A.D., England enacted a watch service ordinance known as a *curfew,* from a French word meaning that all household cooking fires were to be extinguished by a certain nighttime hour. After the fires were extinguished, a watch roved the cities checking for violations or for wildfire. The night watch system was the first step toward fire suppression.

Colonial America faced a constant problem of wildfire. Colonists built their houses roof-thatched in straw and branches close together for safety in case of enemy attack. Sparks and embers from cooking fires frequently sent flames racing through the villages and towns, jumping from house to house. The colonists fought back with leather bucket brigades, long swabs on poles and hooks attached to ropes for pulling down houses to stop the fire's spread.

The London fire of 1666, which consumed thirteen thousand buildings, sparked the invention of the hand-tub water engine. This device on wheels was merely a rectangular water storage box topped by a smaller condensing case and two pistons operated by handles. A bucket brigade at the fire site kept the storage box filled with water. Men taking turns operating the piston handles pumped a short stream of water onto the fire from about twenty-five feet away.

After its second major fire in 1676, Boston ordered

one of these engines, then ordered six additional ones after the fire of 1711 destroyed more than a hundred buildings. Other American cities soon followed Boston's lead. The hand-tub water engine, with refinements, remained the primary firefighting apparatus until around the turn of the nineteenth century.

Ben Franklin might rightfully be called the "Father of American Firefighting." In 1736, in Philadelphia, he founded the first volunteer fire department, the Union Fire Company. The concept of volunteer fire departments spread rapidly after that.

Hook and ladder companies were formed to supplement the engine companies. (The two companies still make up the modern fire department's basic combat arsenal.) Firefighters carried the ladders on foot until they developed a long-wheeled chassis pulled by manpower. This first hook-and-ladder truck also carried picks, axes, leather buckets, whale-oil torches, and fire hats.

The water-tub pumper grew a fire hose around the turn of the nineteenth century and began to draft its own water from a well or cistern or pond, thus eliminating the bucket brigade. Horses instead of men began to pull the apparatus.

Beginning in the early 1800s, cities installed water mains, some aboveground and some below. This rudimentary network following main streets was constructed of pine logs with three- or four-inch holes bored through the centers. There were no hydrants, just a removable wooden plug—hence the term "fire plug"—which allowed firefighters to tap into the main. Fire hydrants were installed later as water pressure increased.

England built the first steam-powered fire engine in 1829. Shortly thereafter, the Latta brothers in Cincinnati constructed the first U.S.-model steamer. Pulled by four horses, it supplied four hoses, each capable of throwing a stream of water three hundred feet. It only took four minutes to get up a head of steam once the boiler was stoked at the fire scene.

Nearly a century later, in the early 1920s, gasoline engines replaced both steam engines and the horses that pulled them. More innovations followed—power-raised aerial ladders, water towers for added water pressure; elevated platforms; fire engines with their own booster tanks capable of shooting 1500 to 2000 gallons of water a minute; fire suppressants and retardants; "slippery" water to reduce friction loss in hoses and thus increase water pressure and volume; dry chemical units; foams, including "light" water and halogen agents.

Cincinnati founded the first salaried professional fire department in 1853. Today there are approximately 300,000 professional firefighters in the United States, plus thousands of other volunteer firemen and firewomen. Additionally, urban areas soon discovered that firefighters were equipped to handle numerous other nonfire emergencies—any catastrophe where life was endangered: tornadoes, hurricanes, floods, mine cave-ins, auto or airplane or train crashes. Fire departments moved into Fire Rescue, providing ambulance services and emergency medical treatment.

While Ben Franklin might be amazed at how his first company of volunteer firemen has grown into a professional army of highly sophisticated and technical fire warriors, there are other things that would probably not surprise him at all. Firefighters, then and now, faced tremendous personal risks. They have become modern society's primary protectors and rescuers, always on call twenty-four hours a day to respond to catastrophe. From the first firefighter to the last, they risk their lives selflessly in the service of others. That would not have surprised Ben Franklin.

Firefighting is the most dangerous profession in America, with a per capita casualty rate even higher than that of policemen. On the average, somewhere in America a firefighter dies in the line of duty every other day. They suffer forty thousand injuries in a typical year. Firefighting is a job in which heroism is part of the job description, in which the elements of human tragedy and disaster present themselves daily.

"Burning infernos," blazing auto accidents, arson murders, attempt suicide rescue, "pot burner fires," lonely old people tripping alarms just to be noticed, heart attacks, industrial accidents, dramatic aerial rescues, a cat in a tree. They are all part of the firefighter's job. In an era in which citizens have grown suspicious of heroes, in which political and business leaders, sports figures, celebrities, and even the clergy have shown themselves in less than an honorable light, the firefighter is still regarded as a hero.

The firefighter may be America's last hero.

3

IF PEOPLE SURVIVED INSIDE THE FONTANA HOTEL, THEY could not survive much longer. Mains drenched a corridor into the hotel lobby; flames howled on either side of it. It was like when Moses parted the Red Sea. Only this time it was water parting a sea of fire. Truckies, one of whose functions was rescue, never entered a fire without the support of a hose. The experienced old-timer Rod Harris led the assault into the inferno, backed by big John Creel and two truckies on one line and by another pumper crew on a second line. They knocked back the fire, maintaining the opening into the smoke.

Smoke so dense it had texture and taste. Fire glowing in the ceiling hurled plaster and burning shards and hot faggots and blazing timbers at the intruders as they steeled themselves and forged their way into the lobby. The environment combined its every hostile breath to evicting the firefighters. The elevator shaft blazed to their right. It blew like a furnace in a tunnel.

Ahead appeared the staircase, indistinct in the smoke, all but invisible until the attackers reached the first steps. Halfway up the stairs lay a woman, collapsed inside her bathrobe like a bundle of dirty clothing, as though washed down with the water cascading from the aerial assault. The flowered print on her robe appeared wilted by the heat and heated water. Creel glimpsed her first. He yelled something at Harris.

The big man took the stairs three at a time and scooped the limp form into his strong arms.

"She's still alive!" he shouted.

John Creel loomed on the stairs, in the boiling smoke, as huge as a Budweiser Clydesdale. He resembled some great futuristic gladiator in his boots and turnouts, helmet and mask. The old woman draping from his great arms appeared no larger than a child, like a little girl fallen asleep on the sofa, whom he was taking to her bed. Even Harris, the old-timer, felt a great welling of pride in just seeing it.

"Get her on out of here!" he shouted, not because Creel needed the encouragement, but instead because shouting made him feel a part of the rescue. He would not have traded the thrill of that moment, that single rescue, to become CEO of General Motors. At that instant he would rather have been exactly where he was than to have been luxuriating in safety in a penthouse suite at the top of the Hilton.

Some experiences just could not be bought.

Indestructible John Creel, head erect, walking tall, literally snatched his survivor from the jaws of hell. There should have been a dramatic orchestra, Harris thought, *something,* as the knight's gallant form vanished into the smoke, following the hose to safety.

Feeling good over one grab didn't cut it. There were other survivors. Maybe.

Most of the first floor had been evacuated out the back door. That left the second and third floors. People remained upstairs whose hopes for life rested on the courage of a small band of helmeted rescuers. Walls alive with fire gnawing at studs and cross members and braces gave Harris an instant's hesitation. No flames as yet showed in the stairwell, but they were on their way. They were coming, following Hades' heated breath.

There was no other escape route except the stairs.

Weighing the risk of being cut off, Harris nonetheless led his band upstairs. A firefighter never runs. Fire school drilled that into every recruit. Glenda Guise sometimes awoke at night with the motto emblazoned before her eyes: *A Firefighter Never Runs.*

Water crashed and splashed in the lobby as the crew on the other line beat at the fire and maintained an open corridor. Jim Barrett, Merker, *somebody,* took Creel's place on Harris's line. Harris operated the nozzle. He kept it going like a machine gun, pumping water into the long second-floor hallway clotted with flame and smoke. Visibility dropped to zero. Fire dripped down the walls.

Miami Beach's aggressive internal assault policy always rode the thin line between courageous and insane.

A sizzling sound coming from a narrow passageway to Harris's left attracted his attention like the burr of a rattlesnake. As he pivoted to confront the new threat, an electrical panel box on the wall exploded in his face like a fireworks display. It ripped his helmet off his head and slammed it against the opposite wall. Harris staggered back, momentarily blinded from the flashbang but grateful for the SCBA lenses that saved his eyes from blast and debris.

He shook off helping hands as his vision returned, although blindness proved little handicap in this smoke. He relinquished the nozzle and groped around until he found his helmet.

"There're people up here...." he muttered, refusing to retreat.

The band fought its way along the fiery hallway, checking rooms for survivors as it went. Truckies kicked in the doors. Harris or Barrett on the nozzle blasted the room with water, shooting high to avoid injury to any inhabitants. The truckies ran in underneath the spraying water, checked underneath the beds and furniture and in the closets and bathrooms, then rushed out again.

"Empty!" they exclaimed each time.

Harris knew people hid up here somewhere, hiding from the flames that found them eventually no matter where they were.

The building started collapsing. Debris pelted the firefighters. It reminded Harris of the self-destruction of the planet Krypton in the Superman movies. He felt that same planet-about-to-go anxiousness. He glanced around, picking out falling embers streaking through the smoke like meteors bombarding a dying planet.

The thought of survivors drove him urgently forward.

Suddenly, reaching with his foot, he stepped into a void. He teetered on the verge of an abyss where the second-floor hallway had collapsed into the first floor. For an instant he gazed straight down into a growling pit of flames.

The truckies grabbed him and yanked him back from certain death. At the same moment, the hallway beyond the abyss lit up. The goddamned hotel was going in just a matter of minutes. Nothing could save it now. Not the entire Miami Beach and Miami fire departments combined. Not even an act of God.

The small band of rescuers hesitated, cut off from

farther advance. Below, the second line maintained the open corridor as an escape route, but above on the roof firefighters scrambled for ladders as their footing weakened and crumbled underneath them. Voices rang out from the lobby, urging the rescuers to return before the hotel became their fiery grave.

Harris and his crew fought on, but a losing battle. Part of the ceiling fell in. The firefighters dodged back, only to be ambushed by a falling wall. Fire rained down onto them. Krypton was about to blow. The entire hotel shook, like a dog shaking off water. It roared and cracked and rumbled.

Retreat for a firefighter, no matter how prudent, always humiliated. Harris looked around and assessed the situation. Although his men were willing to move forward, he would not sacrifice their lives in a foolhardy rescue attempt for survivors who were likely already consumed by the fire. The anguish of that moment of decision ate at him for weeks afterward.

"Let's get the hell out of here!" he decided.

He left victims behind in the fire. He left victims behind while he and his men scurried from the flames like rats burned out of a hayloft. Retreat from the Fontana became more than a retreat; it became a rout.

The stairwell to the lobby had lit up. The rescuers fought their way down the stairs, spot-shooting water at the flames. The rumbling of the dying hotel lent a self-survival urgency to the withdrawal. In the lobby the other main pipe closed in behind the returning firemen and fought a delaying action. Howling flames lapped at the rescuers' heels, drove them coughing and wheezing into the street. The second pumper crew followed just as the entire second floor of the hotel dropped with a terrifying explosion into the first floor.

A runaway hose sailed off the roof, its nozzle spurting water. It twisted and whipped savagely about in the air as it fell, like a snake gone insane with pain. Firefighters still on ladders, ants on twigs weathering a windstorm, held on with death grips as the roof of the Fontana plunged into the conflagration below. It weakened the third floor; it gave way. It fell past the tiers of windows, blazing furiously, gutting the building until nothing remained except walls around an inferno that roared in refueled defiance.

No one inside—not trapped residents, not firefighters—could have lived through it. Lieutenant Rod Harris, face blackened and hair singed, stood in the street gasping for breath and watching with that special awe and respect all firefighters harbor for the power and fury of their enemy. He blinked, surprised that daylight had arrived, and turned away. People had died in there.

Someone placed a gentle hand on his shoulder.

"Not even heroes," a voice said, "can win 'em all."

4

NINE PEOPLE PERISHED IN THE FONTANA HOTEL FIRE. It took days for solemn firefighters led by Lieutenant Vance Irik, fire prevention inspector and investigator for the fire department, to locate all the corpses—blackened, charred, cooked things with the arms and legs burnt off. Grotesque, almost unrecognizable as human remains. Crispy critters, they were called back in Vietnam. Seventeen residents and four firefighters were also injured.

Crews found one elderly woman inside the elevator shaft on the first floor, buried underneath a ton of debris when the floor above collapsed on her. Irik, who pawed through the remains searching for clues, for signs of arson, assumed the woman either tried to take the elevator down or stumbled into the shaft and fell.

The corpse of another old woman lay clutching the charred remains of a box of photographs and other keepsakes she had apparently given her life to save. Irik calculated she died a few feet away from a fire exit, presumably overcome by smoke before the fire reached her and the floors caved in.

Rod Harris, who escaped the hotel seconds before it

succumbed to flames, took each of the nine deaths personally. He looked at some of the bodies where they were recovered. He looked long at them. Then he turned silently and slowly walked away.

Vance Irik could not walk away. As fire investigator, he searched for clues as to how the fire started, knowing that they may have all been consumed. Knowing that arsonists often did their solitary work in public places like hotels.

Twenty-seven percent, over a quarter, of all large-loss fires in the United States are started maliciously. In most criminal actions, it is easy to at least prove that a crime has been committed. The dead body lies there for the police to see; a burglar leaves a broken window behind and takes with him his victim's color TV and VCR. Arson leaves no such obvious modus operandi. Arson is the most difficult crime to prove and to prosecute.

Any evidence of arson is very often consumed by the fire itself, plus, most fire starters are amateurs without criminal records. That makes tracking them and convicting them even more difficult. Eleventh-century England looked upon arson with such dread that arsonists were sentenced to death when apprehended. At the time of Henry II, arsonists were banished from the kingdom or punished by having a hand or a foot cut off. In the United States, first degree arson, the torching of any dwelling or attached building, carries a penalty of up to twenty years in prison. It is murder if anyone dies as a result of arson.

In nearly a decade of investigating fire causes and arson, Vance Irik had seen it all. Motives for arson, he discovered, fell generally into one of four categories, the most prevalent of which was the profit motive.

During economic recessions the owners of small businesses and stores sometimes resorted to fire to get out from underneath mortgages or escape their losses by collecting insurance. Still others torched competitors. Disgruntled partners burned out a business to recoup their investments. In some cities, professional firebugs advertised their services by word of mouth. In a few instances, carpenters and insurance agents were known to go around setting fires in order to generate new business.

The second category was revenge. After a nasty divorce, an ex-husband snuck up one night to his former residence and burned down the house with his ex-wife in it. A man burned down a fastfood restaurant because the chain refused to settle over a fish bone he claimed had gotten stuck in his throat. A fired employee got even by striking a match to his boss's office. Still other revenge motives involved labor disputes, had racial overtones, or were the result of mob actions.

Criminals have been known to use fire to attempt to cover up other crimes, such as murder, embezzlement, or burglary, in a third motive category. The problem with this motive, as most criminals discover to their dismay, is that, rather than cover up the deed, fire usually brings attention to the original crime.

To Irik, the most perplexing of the fire-starter motives, the most difficult to comprehend psychologically, was the fourth category, which falls under the broad heading of pyromania. The true pyromaniac possesses no rational motive for starting fires, other than delusions that center around flame as a sexual or adrenaline stimulant. Irik had discovered through experience that if he arrived early at a suspicious fire, he could walk through the crowd of onlookers and often pick out the pyro. The pyro was always more intense than the other spectators, his eyes bright with excitement. He was either so focused on the fire that he failed to notice anyone around him, or he babbled indiscriminately to whoever would listen. Some pyromaniacs received their sexual jollies from flame; they stood in the crowd watching and masturbating with their hands in their pockets.

While pyromania knows no age limits, sixty to seventy percent of this category are juveniles—and ninety percent of those start out as abused children. Irik had almost memorized the statistics. Kids went out and started fires, usually in areas accessible to the public, subconsciously seeking security or excitement or gratification. It was their unformed way of seeking attention; it was rebellion against authority; it was a substitute sex thrill. Many of them started off by calling in false alarms, then moved on to real fires when the alarm thrill faded. Often they became the "heroes" who warned residents after having set the fire and who "helped" firefighters when they arrived.

Curiously enough, Irik had learned, an abnormally high percentage of adult serial killers started out as juvenile fire starters.

Psychiatrists had studied pyromaniacs intensely, categorizing them, as science will, into various pigeonholes—psychopaths; epileptics who suffered amnesia, became irritable, troublesome, and occasionally firebugs; psychoneurotics under emotional tension; the mentally handicapped; alcoholics or drug addicts; the fetishist who replaced his urge for female clothing with fire; exhibitionists; Peeping Toms; suppressed homosexuals. Not all the people in each pigeonhole started fires, of course, but for those who did, the destructive quality of unleashed flame somehow became a substitute for something they lacked emotionally.

Vance Irik never professed to understand such people. Not being a psychiatrist, at least not in the formal sense, he accepted what he needed to understand about them in order to investigate arson and apprehend suspects. Each time he investigated an arson, he was glad all over again that most arsonists were stupid or careless or even a little crazy. When he arrived at a suspicious fire scene, what frequently confronted him was nothing but a pile of blackened rubble and sometimes a burnt corpse or two. He had several tools he used to help police detectives corral a suspect.

The camera, especially the video, proved important in helping isolate the starting point of the fire, its area of greatest intensity, and the rapidity and manner in which it spread. It pinpointed smoke and its concentrations, its color and activity. This, in addition to statements from other firefighters and witnesses about whether the doors and windows were locked or unlocked, opened or closed, and such things as odors, helped him determine where and when the flame started, how it started, and with which fuels. The fire chief and police would be asking him to prove the corpus delecti of arson—that the blaze was started by criminal design, eliminating any accidental sources, and that it had been started by a particular individual using a particular incendiary at a particular location within the building.

But perhaps a fire detective's most important tool was his shovel. At best, arsonists were seldom original. That meant they usually went for the more simple incendiaries, such as gasoline, crumpled papers, matches and kindling. Irik sifted through the debris with his shovel, looking for evidence, taking still photographs as he went along that could be presented in a later trial to prove arson.

Lieutenant Irik and his shovel had become familiar at fires with questionable antecedents. Because the Fontana Hotel was almost completely destroyed, the cause of the fire would probably have to go down officially as a fire of "undetermined origin." Sometimes Irik had to accept it—fires that killed might be an act of God or an act of man, and he would never know which.

5

ONE OF THE STRANGEST ARSON CASES VANCE IRIK HAD EVER worked was that of the nude man at the hotel fire. Barrett and Merker and the other firefighters ribbed him mercilessly about it.

"Vance, as a trained investigator, when was it exactly that you noticed the guy wasn't wearing nothing but a smile?" Stu Merker probed, his eyes twinkling as mischievously as those of a squirrel.

Irik hadn't even needed his shovel to uncover the mystery of the nude arsonist. The whole thing, as Merker put it, had been *bared* for him.

It was a nine P.M. Station 2 pushout to a fleabag hotel on Collins Avenue, a bed fire on the second floor.

It hadn't been much of a fire. The bed was a pile of charred mattress, and the walls were peeled and blackened, but flames had not spread beyond the one room. Some guy not wearing a stitch of clothing had run up and down the hallways alerting the residents.

Lieutenant Rod Harris left a couple of firefighters and the pumper behind for overhaul. They were still upstairs at the fire site when Irik arrived, summoned because the blaze was a suspected arson. The police had cleared the area so people could go about their business, and Collins opened again to a steady stream of tourist traffic. The pumper sat parked across the street in front of a closed clothing store. Spotting movement on it, Irik strolled over to get the firefighters' story about the arson.

He approached the fire engine, then froze in his tracks. He blinked and looked around. The only person within sight was the guy on the pumper—a skinny man who looked old, so much older than any man could be, and haggard, as if he had just constructed the world with his own hands, or had just destroyed it. His eyes shone wide and as dark as spots of used oil.

He stood with his nudity partly concealed by the fire engine from the Collins Avenue traffic. Sand, as from a grave, coated his frail body. He turned to stare down at Irik. Irik took an involuntary step back, momentarily speechless. How did you approach such a man?

"I'm so terribly sorry," the nude began, his lips quivering. "I didn't mean to hurt nobody, but I didn't have no choice."

He sounded harmless enough. Irik looked around for a cop. Seeing none, he edged closer to keep the guy talking and prevent his fleeing if he suddenly took the urge. And there the tall uniformed fireman stood on busy Collins Avenue, holding what appeared to be a normal conversation with a jaybird-naked man standing on the fireman's fire engine. Someone in a passing car pointed out his window and laughed.

"I like firefighters," the nude continued in a diminished voice. "I wouldn't want to see any of you get hurt."

What the hell. Irik kept the conversation going. The naked man leaned his elbow on the top handrail. His eyes shifted nervously as, in an apparent need to justify himself, he explained how he had intended to burn down the hotel and everyone in it, himself included.

"The voices told me what to do," he explained, his voice unnaturally calm, considering he could have burned alive a hundred people or more. "The voices told me to cleanse myself with fire. But first I had to urinate on myself three times. Then I jerked off and took some cum and made a cross on my forehead. That's what the voices told me to do. I knew I had to die too in order to cleanse myself with fire. So I took off the rest of my clothes and lay down on the bed and struck a match to it."

Irik stared. Sometimes, such as now, he considered returning to an engine or ladder company. Firefighters saved lives and property. Like the police, an arson investigator witnessed so much of human nature's dark side that he grew cynical and suspicious.

"It got so hot in bed," the nude continued, "that I changed my mind. I couldn't do it."

He jumped up, warned his neighbors, and fled the hotel while his bed blazed. He said the voices ordered him to drown himself instead. He raced to the nearby beach, where he waded into the dark surf and tried to sink. He kept coming up to gasp for breath. Finally, waves washed him onto the sand, where he lay for a while weeping because he was such a failure.

Some of the displaced residents from the hotel gathered on the sidewalk in front to wait for the smoke to clear. They made threatening gestures at the nude and shouted what they would do to him if he dared return. The nude glanced lazily across the street at them.

"I'm glad the hotel didn't burn," he murmured. He looked back at Irik with his wild, lonely eyes.

"Do you think the goddamned voices lied to me?" he asked.

Something in the guy's dark eyes reminded Irik of little Manny, something pleading and afraid. Irik formed a mental picture of the guy naked and huddled alone in the darkness of a cave or a sewer or something. Manny had that same kind of afraid, alone quality.

6

EACH YEAR AROUND CHRISTMASTIME SOMEONE ALWAYS brought up Manny's name. Ramirez or Barrett or Captain Garcia stopped whatever they were doing for a moment, looked around expectantly, and then murmured, "Wonder whatever happened to little Manny?" The firefighters of Firehouse 2 often associated Manny and Christmas, for they remembered that it must have been Christmas that year when Manny, about ten or eleven at the time, started hanging around the firehouse.

Manny was a skinny kid, small for his age, as if he hadn't started growing, with brown eyes like giant buttons on a teddy bear. He brought with him an afraid, alone quality. Already he had the face of a grown man who had labored in coal mines.

He came to shoot hoops with the firefighters on their court and to watch them collect canned goods for poor families and wrap Christmas packages for needy children. Christmas was not only the busiest time of the year for both fire suppression and Rescue, it was also the traditional time when firefighters conducted drives to bring Christmas to people who might otherwise not be able to celebrate it.

"Manny is about the neediest kid I know personally," Rod Harris commented in his deep voice, which somehow grew softer when he spoke of the kid.

Christmas meant Manny from the first time he appeared. The firefighters always had something for him underneath their tree in the ready room. Station 2 became Manny's second home, the firemen his family. He was always there, hanging around, wearing an old fire helmet, shooting hoops, watching his heroes with that particular adoration of a small boy who, having nothing else, finds *something*. Someone found the kid sleeping in the pumper one night.

"Kid, where's your dad?" someone asked him.

"I don't know. I hope the sum'bitch is dead. He won't beat me no more."

"How about your mom?"

Manny gave a casual shrug and walked off. He wouldn't answer that question. The firefighters later discovered that his mother was a hooker working the hotels and joints on Collins Avenue. The streets were mostly little Manny's home. The streets and Firehouse 2.

The boy seemed especially intrigued with Vance Irik. "Lieutenant Irik, whatta you do?" he had asked. "You don't jump on the truck with the others."

"Mostly, I enforce the fire codes and investigate suspicious fires," Irik explained. "Arson," the lieutenant explained, when Manny looked puzzled. "Arson is where someone deliberately sets a fire."

"Why would somebody do that?" the boy demanded.

For several years after that Christmas when Manny first showed up at the firehouse, he was always at the station to greet firefighters returning from a run.

"Tell me what happened," he insisted. "Tell me what it was like."

After an elderly woman knocked over a candle and

set her condo afire, Manny asked, "How can she be so stupid to set fire to her own house?"

"People make mistakes," Irik said. "It was an accident."

"Fires are not always caused by accident?" Manny asked.

"Not always."

"But the firefighters come?" Manny asked. "If it's an accident or not, the firefighters still always come to put it out?"

"Yes. They always come."

Manny ran off to play catch or something with Ed Delfaverro.

Then, one day, Manny was gone. Christmas that year around Station 2 was like the only child had grown up and left home. It would be another Christmas in the future before Vance Irik encountered Manny again. The first thing he would notice about this different, grown-up Manny were the eyes. They had the pleading and afraid look in them of a naked man huddled alone in the darkness of a cave or a sewer.

7

THERE ARE LEGENDS IN FIREFIGHTING JUST AS THERE ARE legends in every other occupation fraught with danger and adventure. The timber industry has Paul Bunyon and Babe the Blue Ox; fighter pilots brandish the exploits of the Red Baron; while the army takes Sergeant York and Korea's Lost Patrol to mythological proportions. Firefighters look to Ben Franklin and his first fire department for inspiration, as well as to a long list of famous Americans who served as firefighters—George Washington, Thomas Jefferson, Paul Revere, John Hancock, Benedict Arnold . . .

Rather than elevating single persons to legendary status, however, firefighters commonly endow the nation's big fires with myth—the Great Chicago Fire of 1871, which supposedly started when Mrs. O'Leary's cow kicked over a lantern; the Brooklyn Theatre Fire of 1876, in which nearly three hundred people were killed; and even fires of the American Revolution, such as when the British set blaze to four hundred dwellings in Charleston, across the river from Boston, shortly before the battle of Bunker Hill in 1775. Individual fire departments, even down to firehouses, nurture their own fiery legends.

Just as Homer's tales were passed down from

generation to generation by word of mouth around campfires, firefighters retell and relive past adventures and enhance them until everyone can tell the stories, though few remember the individuals involved. Christmas especially offers the opportunity for fire departments to keep their traditions alive through the art of storytelling. For a month or more around the holidays, Firehouse 2 on Miami Beach hosted a steady stream of traffic in and out—spouses, children, girlfriends, parents, friends, and retired smoke eaters like wiry Gene Spear, who revived the old stories and added to them.

Firefighters might be considered blue collar America at its best. While politicians give lip service to family values and operate privately on a much lower level, firefighters *live* the American values. They still possess a sense of community, in that they are known and appreciated within their fire districts by everyone —from kids like Manny, who stop by to shoot hoops, to the district representative. They try to live up to an image. Their family starts with spouse and children and extends to include the firefighters at their own firehouse. Spiraling out from there, family encompasses the community, the entire fire department, and, finally, all firefighters. In the early history of fire departments, a community's firehouse also served as its social hub, where men gathered to talk politics and women held quilting bees or organized fund-raisers for the needy. Station 2 carried on that social tradition, at least during the holidays.

Sometimes raucous, opinionated, occasionally prejudiced, often stubborn, firefighters nonetheless are as ready to risk their lives to save others as they are to listen to or tell a good story. Everyone from the fire chief to the newest probie looks forward to the holidays, in spite of the Silly Season's thirty alarms or more a shift. The mess hall table is always laden with cookies and cakes and other treats. Some of the more accomplished station house chefs like Jim Barrett or David Mogen, or at times Glenda Guise, whip up special meals that turn the firefighters into potential replacements for the jowly Hog's Head Chief that hangs on the ready room wall. As well as a time for work, it is also a season for visiting, reminiscing, and for fire station tales.

Gene Spear had hung up his turnouts the year before to retire to Florida's Keys. "When I joined the fire department," he announced to an audience that included a half-dozen wide-eyed children and a gray-haired neighbor lady who dropped by with fudge, "the only formal training I underwent was an American Red Cross first aid class."

"The continuing story of America's heroes," quipped lively Stu Merker, whose wit made him the station's mischievous younger brother. "Gene, tell us how you drove the horse-drawn fire wagon through three feet of Miami Beach snow—"

"It was only two feet," Spear shot back, laughing.

Spear, who had given twenty-five years to the job, referred to himself as one of the original smoke eaters from back before modern equipment and training made fighting fires so easy. Back when men were *men*.

"I was hired on Friday," Spear liked to explain, "and reported to the station house on Monday. It wasn't like our probies now, who go through ten weeks of school just to learn that a fire burns if you touch it. On Monday and Tuesday I worked with the ladders. On Wednesday and Thursday I learned how to use the hoses. On Friday the chief told me to buy my boots and suspenders and pick up my coat and helmet—I was a firefighter. He said to report Monday for assignment. The first time I found myself in a big fire, I asked myself, 'What the hell am I doing here?'"

"When did you stop asking yourself that?" Neal Chapman asked.

Spear laughed. "When I retired."

Even into the 1960s and 1970s, many metropolitan firefighters in the United States received all or most of their training on the job. As with Spear, somebody showed them where the hoses and ladders were kept, issued them turnouts, and presto, the next thing they knew, they were battling fire. Local legend has it that when Ocean Beach became the City of Miami Beach during the Florida land boom of the 1920s and the city bought its first fire engine, J. S. Stephenson was appointed fire chief because he was the only man who knew how to use the pumper.

Since then the Miami Beach Fire Department has served its community through times that molded and tempered it. Old-timers like Gene Spear and Captain Jim Reilly have lived much of the department's history as Miami Beach grew and strained at its island seams on its way to becoming a tourist mecca. A 1926 hurricane flattened nearly every structure on the island, while Hurricane Andrew in 1992 missed it by about ten miles and destroyed Homestead to the south. Over the years, political conventions, unrest over Vietnam, racial tensions, hard economic times, have all taken their toll on the city, as they have taken their toll on the nation. The fire department and other public services responded to the challenges and improved as a result of them. Spear and Reilly had witnessed the 1960s protests, with their burned cars and bonfires and arsoned buildings; the demonstrations that incited responses in tear gas and gunfire. The Miami Beach Fire Department hurled itself into the modern age, leading the way for other departments by innovations in Fire Rescue and Fire SWAT.

The current fire chief, Brainaird Dorris, continued the work of his predecessor, Bud Goltzene, in more sophisticated training for firefighters and in equipment upgrading. Firefighters boasted they could respond to a fire anywhere on the island in less than three minutes, with the correct force to handle a burning oil barrel or a towering inferno. That was critical on an island where buildings stood elbow to elbow and fire could skip from one to the other like a child going to the store for candy.

Although the fire department was not large as far as metropolitan forces were concerned—with only slightly over two hundred troops manning four station houses on the tiny, crowded island—it was one of the busiest in the United States and was often used as a model by other cities looking to upgrade their own fire forces. On Miami Beach the fire chief no longer acquired his job because he was the only one who knew how to use the pumper.

The troops preserved the changes and the history of their fire department through the oral tradition of storytelling. Stories illustrated lessons learned and were therefore educational. Every probie in fire school, for example, knew the tale of Lenny Rubin.

Fire broke out in one of the smaller hotels off Collins Avenue during a minor hurricane. Somehow, Rubin got separated from his crew in the blaze. He found himself momentarily alone in the flooded basement where he slipped and fell in the thick smoke and apparently knocked himself unconscious. He drowned in the shallow water.

"The lesson is clear," fire instructors pointed out. "Don't get separated when you can be picked off one by one; and if you do get separated, make sure you know where the lifeline is—the hose—so you can follow it back out of the fire."

"There are *old* firefighters, and there are *bold* firefighters," Gene Spear always said, grinning, "but there ain't many *old, bold* firefighters."

Firefighter legends also were used to enhance and pass along acts of bravery and resourcefulness, as an inspiration to the probies and younger firefighters, much in the same way that the old Uncle Remus tales used Brer Rabbit as an example of cleverness and imagination triumphing over brute strength in the form of Brer Fox or Brer Bear.

"'Skin me, Brer Fox,' sez Brer Rabbit, sezee, 'snatch out my eyeballs, t'ar out my years by de roots, en cut off my legs,' sezee, 'but do please, Brer Fox, don't fling me in dat brier patch.'"

Gene Spear leaned forward, ready with his story. The name of the principal hero had long been lost, but that didn't matter. Children scooted closer. Stu Merker, Jim Barrett, Glenda Guise, and the other firefighters found chairs.

8

IT WAS ONE OF THOSE FIRES, SPEAR SAID, THAT COULD really turn a day's schedule upside down. It was a kind of running joke that a working fire could be a problem, because it screwed up the regular work schedule. Most fire departments worked a three-shift schedule, with each shift at the firehouse for twenty-four hours on and then off for forty-eight. Each shift became a kind of separate family, with strong ties and loyalties.

A shift on Miami Beach began at seven in the morning. The first thing firefighters did was check out the apparatus and equipment to make sure everything was functional—oil, gas, lights, siren, SCBAs, hoses. Housekeeping usually followed—mopping, cleaning, buffing, whatever—while the shift officer caught up on paperwork and planned the afternoon.

Afternoon schedules included building and district tours to prefire plan: "What happens if this high-rise condo caught fire on the twelfth floor and we had to evacuate survivors?" That kind of thing, followed by business code inspections, drills and classes, and maintenance around the station. The evenings were generally free for the firefighters to work out in the exercise room, play basketball or catch, or whatever. After dinner they watched TV, read, studied for promotions, or sat outside in the Florida weather and exchanged firehouse stories or gossip.

Of course, all that was dependent upon alarms. The schedule went all to hell when an office building on Collins Avenue ignited, pushing out a basic attack team of two engine companies, a ladder company, and a Fire Rescue team.

Our hero, Spear said, was working Engine 2. Engine 2 drew first-in on an interior blitz after the truckies opened up the building.

Firefighters normally work on a "team" or "buddy" system in which they look out for each other to prevent their getting isolated or trapped by flames. Our hero, however, somehow became separated from his crew and hose line. He stepped aside momentarily to hack at and scatter a pile of debris with his ax. He followed flames through a doorway into an office. He was still a probie, so the story went, and although, as it turned out, a particularly inventive probie, he was still a probie, and probies sometimes did foolish things.

Tapping the walls with his gloves, he edged through the thick smoke toward the hallway. To his surprise, he found his path blocked by a wall of flames that cut him off from the door. The flames bent and swayed and cracked like bullwhips. They were really cooking, turning the office into a hostile environment filled with heat and toxic smoke. Blinded by the smoke, unarmed, he retreated. He stumbled over a desk. The wall farthest from the hallway door and the roaring fire stopped him. The heat was intense. The most heat a man could withstand was about 300° Fahrenheit and still breathe, even from a bottle. The air in the probie's bottle already threatened to scorch his lungs.

He felt his way along the wall, searching for another exit. This was an inside office; there were no windows to the outside. Fighting panic, he came to a door that he found opened into an adjacent office. He slammed the door behind him to starve the fire next door of oxygen. No yellow tongues of flame had yet crept into this office, although the room was heating up fast.

Smoke remained heavy, churning. He explored the room with his hands. To his dismay, he quickly discovered that this must have been an executive's inner office, while the other office was the reception area. There were only two doors to the room, and again no windows. One door led into the roaring inferno next door; the other opened into a small windowless bathroom.

Heart thumping wildly, he backed against the wall farthest from the fire. At the same moment he realized he was trapped, another more urgent problem confronted him. His air bottle when fresh contained thirty minutes' breathing. He sucked the last of the stale air from the bottom of the tank and then found himself sucking on a vacuum. Still fighting panic, he ripped his air hose from the tank and dropped to the floor to taste the fresher air next to the carpet. It filled his heaving lungs with taint and heat.

Hacking viciously, eyes and nose running, he opened his turnout coat and ripped pieces of cloth from his shirt, which he wrapped around the end of his tank hose to filter the air. The air still tasted thick and bitter, but he tried to comfort himself with the knowledge that his crew must have missed him by now. Surely they would find him if he could hang on for a few more minutes.

Would he bet his life on that?

He couldn't just lie there and wait. The office was starting to really roll. South Florida pine walls cracked like bursts of machine gun fire as flames tongued through the wall from the adjoining office. The office lit up eerily in the smoke. The probie had a desperate feeling that the fire knew it had him trapped and was laughing at him, teasing him like a cat with a cornered mouse.

He understood clearly now why livestock trapped in a burning barn often lost all control and stampeded directly into the flames to their deaths. It took all his willpower not to charge into the fire toward the only known exit. Just go mad and get it over with.

He thought he might have prayed then, a little, but maybe he just closed his eyes against the stinging heat. His eyes felt like grapes about to pop from their skin. As there were no atheists in foxholes, there were likewise no atheists trapped in a fire. Burning to death, he thought, must be the most horrible of all deaths.

Unwilling to simply lie there and succumb to the smoke, he crawled across the carpet toward the bathroom door. Patches of flame gnawed at the carpet. Gasping for what little good air remained made him dizzy. He thought he might pass out. Fainting from lack of oxygen before the flames reached his body was a more merciful way to go.

He paused at the bathroom doorway to peer inside through the boiling smoke with cold dread. Once he entered, there was no return. His comrades might easily overlook the bathroom, and he inside it.

But what other choices did he have? Flames steadily ate their way toward him. He crawled into the bathroom as though entering his own tomb. He slammed the door behind him and leaned heavily against it, as though he might stop the fire by sheer force of will. It was as dark inside the tiny room as inside a closed coffin. The fire outside devoured its relentless way toward him with the sound of warrior ants chewing down a forest. The door would not stop the march for long. It was nothing but fuel.

While fire school tried to prepare prospective firefighters for such emergencies, no probie ever understood the sheer stark terror of the enemy's power until he faced it for himself, alone. There was something almost genetic about man's fear of fire. That fear stretched back into the generations to cave people fleeing blindly from forest fires racing through the trees faster than a man could run. The probie's heart pounded so fiercely in his chest that he thought it would break his ribs. He almost collapsed from fright and heat and lack of air. His head spun.

He had to think. Had to. Lack of oxygen quickly robbed the brain of its ability to reason. *Think, goddamnit, think!*

Either that or die like a rat trapped in a microwave oven.

The probie felt himself starting to drift. It was almost like he was floating. He shook his head hard to clear it. He heard himself panting for breath. It occurred to him that when the rescue crew reached him, it would find him belly down on the floor where he had sucked in his last air. Dead.

Wait a minute. A thought struck him with such sudden clarity that he almost laughed aloud. Maybe he had a chance after all. Either that or he was hallucinating.

Muttering, "Not yet, damn you. Not yet," he clambered to his feet and groped his way around the wall. He found the toilet stool and the sink. He remembered the old high school physics trick in which if an open bottle of water were turned upside down quickly, the water remained in the bottle because air could not get inside to replace it. It was similar to the old days when juice came in cans. Unless you opened a second air-vent hole, the juice would only trickle out. Same thing with plumbing. Plumbing was always vented to the outside of a building because water in a pipe would not run freely unless air reached it.

With diminishing strength, knowing this was his last chance, his *only* chance, the young firefighter hacked at the wall with his ax. Plaster flew. He felt the ax break through the wall. He enlarged the hole until it was big enough for his hand. He felt around inside the wall as far as the crook of his elbow permitted.

Nothing.

He moved over and hacked out another hole. Again he felt inside.

Again nothing.

He paused to cough corruption from his lungs. He hadn't much strength left. Not much time. Suffering from vertigo and claustrophobia, he braced himself against the sink to stop his spinning. He felt around on the wall in the smoke until he found the two fresh holes in the wall. He shifted two feet to the left of the first he had made. From beyond the door came the cracking and slavering of the hungry flames as they rapidly lapped up the remaining oxygen.

The probie thought he heard someone yell his name, but he couldn't be sure. Symptoms of poisoning by gas included dizziness, dryness of the membranes, intestinal disturbances, increased heartbeat, and delirium. People heard things that weren't there when they were delirious.

The firefighter tore at the wall with his ax. He was so weak from smoke inhalation by this time that the ax blade bounced back at him as though off metal. He kept at it desperately, chipping away at the wall. Later he explained that the thought of never enjoying another root-beer float on a hot summer's day, of no more moonlight swims in the ocean, of never kissing his girlfriend again, drove him to the lengths of his endurance. And then beyond.

Finally, he hacked a hole in the wall, but it wasn't big enough for his hand. He kept chipping at it. He turned on the water in the sink and let it fill the sink and splash onto the floor. That might keep back the flames for a while, but unless he found air quickly, the matter of the flames reaching him became a moot consideration. He would be long dead of asphyxiation by the time the fire gnawed through the door.

He reached into the wall through his latest hole—and tears of joy filled his eyes.

When the rescue crew located him a few minutes later, he remained alive only because of his quick thinking. There he was in the bathroom, so the legend went, with the hose of his SCBA thrust into a rent he had made in the plumbing ventilation pipe that led to fresh air on the roof of the building.

He grinned broadly. "What kept you guys so long?" he demanded of his surprised comrades, who had expected the worst.

Retired firefighter Gene Spear sat back in his chair, clearly pleased with what the story said about the character of the people who fought fires. Almost at the exact moment he finished, as though by dramatic cue, the shrill of the alarm rang through the firehouse. Firefighters sprang to their feet and jerked on turnouts as they sprinted for the engine.

"Daddy? Daddy, can I go too?" a young son shouted after his father.

Gene Spear and the little boy watched the fire engine leave the station. Man and boy wore much the same expression.

9

OTTO RAMIREZ, WORKING RESCUE 2, SAT ON WATCH IN THE ready room, idly playing with a glass paperweight filled with a winter's farm scene. He shook it to make it snow. Late afternoon shadowed out the room as the young, powerful-looking U.S. Air Force veteran with the amazing blue eyes stared into the make-believe snow.

He was tired and his shift wasn't even half over. Since noon Rescue 2 had made ten runs. A couple of them had been false alarms, and then a drunk fell off his bar stool. Two cars collided; another took a swim off the Tuttle Causeway bridge, which linked the island to Miami and the mainland. Heart attacks followed by a construction accident. Then an elderly guy on South Beach stuck his penis into a Regina electric broom—"More suck for the buck"—and got it too near the blades.

That prompted a fresh assault on heavily mustached David Mogen. Mogen was about to marry his Jewish girlfriend. The other firefighters ribbed him about having to undergo a circumcision in order to convert. They came up with all kinds of inventive methods of accomplishing this. Stu Merker immediately decided an electric broom was the answer.

"It's over just like that," the impish little man quipped, snapping his fingers. "Naturally, you don't want to get too much of it stuck in there—or there goes the honeymoon."

Laughter and loud talking from the garage below where the men cleaned equipment and teased Mogen jogged Ramirez momentarily out of his reveries. But he was too tired to join the others. During the holidays, Rescue rode ten calls to every fire call, sometimes more. In 1978 then-Fire Chief Bud Goltzene set up a policy of rotating firefighters between Rescue and fire suppression to keep down paramedic burnout. That meant all firefighters, in addition to fighting fires, underwent emergency medical technician training and rotated tours on Rescue. The policy had endured.

Unlike some firefighters who disliked Rescue and its dirty, dangerous, though sometimes funny work, Ramirez often volunteered for additional tours. He was attending school off-duty to become a nurse. Rescue gave him practical experience in emergency medicine. Although the gore and the senseless tragedies, the sick things like knives and guns day in and day out, sometimes got to him, he found great personal reward in helping the victims of his densely populated city. Saving lives gave him great satisfaction.

Only thing was, when he wasn't at the station house or at school, he worked an additional job as a dive technician treating scuba-diving ailments. He was often so beat after a twenty-four-hour shift on Rescue that he simply hurried home and crashed before he had to get up and go again. Sometimes he felt his wife and two kids were strangers living with him underneath the same roof.

"Do I know you?" his wife asked, not quite joking. "Your name's Ramirez, isn't it?"

Years before, New York City discovered that the fire department was the perfect nucleus around which to mobilize better emergency service. While the U.S. Highway Safety Act dealing with the issue of increasing traffic fatalities sparked the organization of and funding for Fire Rescue, and although New York and other cities were already experimenting with the concept of twenty-four-hour emergency service, it was not until 1973 that the State of Florida became the first state to legalize a paramedic corps operated by fire departments.

Prior to that, funeral parlors commonly ran ambulance services. The hearses were operated by men whose medical skills were limited to placing victims on stretchers and roaring away with them to the nearest hospital. Some patients simply got up from dying and *walked* to the nearest hospital rather than ride in a black hearse smelling of formaldehyde and roses.

Fire Rescue on Miami Beach actually began in 1965 when the city acquired its first rescue unit—a 1950s-model station wagon known as the "Beast." Like their predecessors in hearses, firemen were poorly qualified to do anything medical other than transport the victim. The Beast's major piece of medical equipment proved to be an ashtray full of dimes so the Rescue men could call for help if they found something more serious than a sunburn or a cut finger.

After 1973, Miami Beach hired medical doctors to ride Rescue with firefighter drivers and crew. In 1986 all medical doctors were finally replaced by skilled and highly trained fire department paramedics capable of handling virtually any emergency from a traffic pileup to a heart attack. In addition to fighting fires, Fire Rescue men became emergency medical experts, rescue specialists, social workers, and masters in public relations. They delivered babies, pumped stomachs, dived on sunken boats, extricated auto crash victims from mangled steel. Their rescue trucks came equipped with sophisticated equipment: drugs, oxygen, IVs, lifepacks used to defibrillate and service heart patients, telemetry units that enabled hospitals to monitor the conditions of victims en route to the hospital.

The trucks were mini-hospital emergency rooms manned by a crew of three firefighter lifesavers. In such troubled times, each unit also came equipped with bullet-proof vests. Some of the rescuemen even went armed. Fire Lieutenant Ed Delfaverro headed a five-man Fire SWAT team of medics created to support police during riots and commando operations.

Wasted humans produced through poverty, population pressure, and personal isolation turned cities into battlegrounds. Some people knew a city by its bookstores and restaurants and nightclubs. EMTs who, because of tragedy, stepped into people's lives for a moment and then out again, knew a city by the bloodstains on its pavement.

Yet some of Otto Ramirez's warmest memories were of victims who returned to have photographs taken next to the fire paramedic whom they felt had saved their lives. In New York City or in Los Angeles or in Miami Beach, communities invariably looked upon Fire Rescue men as genuine heroes.

"It's a job," Ramirez liked to say modestly. "You can't save 'em all, but you can save some of 'em."

It was more than a job, and he knew it.

He sat relaxed in the ready room playing with his paperweight, watching it snow. He looked up as Vance Irik, the fire investigator, entered after looking over a suspicious hotel fire. Miami Beach had suffered a number of recent hotel arsons. Irik's face was blackened from digging through fire rubble searching for clues. Irik nodded, in a hurry, and went on through the ready room and out.

"Know what I would like to see in Florida?" Ramirez mused.

Chapman paused. "Santa Claus?"

"Naw. Santa Claus would get mugged." He held up the paperweight. "Snow. Christmas just isn't Christmas without snow."

South Florida was having a "cold wave" in December; it was seventy degrees outside.

"I'll take the sunshine," Chapman decided. "You can pretend it's snowing."

Ramirez frowned. "Neal, you remember that kid Manny? He used to be around every Christmas. Whatever happened to him?"

"Manny," Chapman repeated. "Yeah. Poor little kid . . ."

The alarm interrupted. The radio toned out a *"Rescue Two, one ill person and a Code 45 at . . ."*

A Code 45 was a DOA—Dead on Arrival.

10

AS MUCH AS IT WAS AN ISLAND PLAYPEN FOR TOURISTS, Miami Beach was a community of singles and of strangers. Rootless young "snowbirds" fled the northern winters. Some were penniless and they slept on the beaches and in the parks. The homeless drifted south too: the mentals, the dopers, the alkies. They shambled around in clothing rotting at the crotches and underpits; even the jails in Florida were warmer than the jails in Boston or New York.

Retirees flocked south to Miami Beach too, in lesser numbers than in years past, but they still came. It was sometimes remarked that the Beach was like the elephants' graveyard where the old northerners came to die. Corpses often lay in rented rooms for days before someone noticed the stench.

Rescue 2's Code 45 address led Ramirez and his two-man crew to an ancient retirement hotel on South Beach, where retirement enclaves were juxtaposed among yuppie condos and high rises. Ramirez cast a weary glance at the gutter pipes hanging desolately from the roof of the two-story rattrap, at scabs of white paint peeling off the building and hanging in the tropical sun. Old people were always falling in the bathtub or getting sick or breaking a hip or something.

They crawled painfully to the telephone and dialed 911 to summon either the police or Fire Rescue. For most of them, having long outlived friends and relatives, 911 was all they had.

When Ramirez knocked on a ground-floor door, an elderly female voice responded so weakly that it might have belonged to a sick kitten. "Hello? Hello? Please, please come in."

As soon as the Rescue man opened the door, he wished he were somewhere else. He dodged back, ducking like a prizefighter, as a hail cloud of greenbead flies enveloped him. The stench of putrefying flesh assaulted him like a wall dripping with gore and pus. It was like something described in a good horror book, like the fly scene in *The Amityville Horror*.

With a burst of willpower, Ramirez held his ground and peered into the darkened room, blinking. All the shades appeared pulled, the windows closed against the intrusion of any fresh air. The other two Rescue men, less experienced in such things, blanched and stumbled outside. Ramirez heard one of them retching up the two coneys with onions, relish, and melted cheese he had had for lunch.

You never forgot the smell of a decaying human body after the first time.

"Jesus God," Ramirez breathed. He tried not to breathe.

"I'm terribly sorry, young man," squeaked the voice. Ramirez squinted into the dimness. Gradually his eyes grew accustomed to the lack of light and picked out a day bed in the room to his left. He looked hard and saw the yellowed head of what appeared to be a skeleton stuck up out of the sunken middle of the mattress.

The skeleton continued to speak in a voice like a shriek diminished: "It's my husband who needs you."

Not comprehending: "Your husband?"

"Of course. He fell down and can't get up."

Judging from the stench, there was a reason why he couldn't get up.

Flies clouded around the skeleton's hand as it rose weakly out of the mattress. It indicated a direction opposite the bed. Ramirez still hadn't moved past the doorway. His eyes shifted. He swallowed his gorge when he spotted two bare feet sticking out from underneath a coffee table. The feet were bloated, black things that resembled inflated inner tubes. Rice crawled all over the feet.

"Jesus God."

Ramirez felt as though he had been transported directly into a scene from HBO's "Tales from the Crypt." He failed to notice his companions returning with Vicks salve smeared underneath their noses to help kill the stink. They stood behind him just outside the doorway, no more eager than he to venture into the room.

It had to be done, but not right away. It took some getting used to at first, even for Ramirez, who had experienced such things before. He didn't need to examine the old man underneath the coffee table.

"Ma'am," he began as compassionately as possible under the circumstances. "Your husband has expired. I'm sorry."

The squeaking voice took on an edge: "Nonsense, young man. I've been talking to him."

Ramirez took another look. He stepped into the room. The old woman had been in here so long with the corpse of her husband that she really had gone batty.

Someone nudged Ramirez from behind. He flinched. "Otto," the Rescue man whispered hoarsely. "It moved."

"It *what?*"

"It *moved.*"

Ramirez stared. "It's *not* moving."

The voice that suddenly crackled from underneath the coffee table sent Ramirez back the step he had taken into the room. The three rescuemen huddled in the doorway, stunned. The voice sounded surprisingly strong, considering it originated from a dead man.

"I could use a bit of a hand getting up from here," the voice said.

"I told you he was alive," cackled the skeleton in the bed.

The paramedics willfully overcame their revulsion in the face of a true need for help. But what they saw when they lifted the coffee table and gently turned the old man onto his back was something that haunted their nights for months afterward. He had been lying facedown in a bed of rotted flesh and human excrement alive with maggots. Flies swarmed as around a road-kill carcass. They torpedoed against the Rescue men's faces, as though fighting off interlopers.

The old man was literally being eaten alive.

"It's hell being old and helpless," he murmured apologetically.

His wife wasn't in much better condition. She weighed about eighty pounds and she suffered a fractured hip. Sometime during the past week she had contracted diarrhea. The weight of her own excrement and the resulting rotting flesh on her backside had compressed a slick nasty trough in the middle of the bed. The paramedics called for ambulances rather than transport the frail creatures in the rough-riding

"Five?"

"Three to hold me down, one to hold it up, and the fifth on the hacksaw."

"Christ, Mogen. I think he took too much of it. Two more inches off and you'd have been a princess."

Lieutenant Vance Irik sometimes missed life in the firehouse where he had spent so many years before moving up to fire investigator. His office was in the administration building, separated and isolated from Station 2. During the Christmas season, he was as busy as the other firefighters. First, there was the increase in accidental fires. Often, this time of the year, a suspicious fire could be traced back to Christmas tree lights whose cords and plugs and bulbs shorted out. Irik always experienced a great sadness arriving at a fire to find a weeping family wrapped in bedclothing huddled outside the blackened hulk of what had once been their home. The loss seemed especially catastrophic at Christmas.

Irik and his shovel kept busy digging through ashes and debris. Some of the cases were as simple as the house fire that trapped twenty dogs inside and burned them to death. Sifting through the ruins, the investigator discovered what was left of books piled on top of a heating pad plugged into the wall.

The discovery led him to a forty-year-old retarded woman, the daughter of the homeowners and caretaker of the pack of now-dead dogs. She knelt on the front lawn, weeping miserably over the body of one of her scorched pooches. After a few minutes talking with her, Irik got up and shook his head sadly and walked off. She could not comprehend the relationship between the fire and the pile of books she had stacked on top of a hot heating pad.

Other cases were criminally devious, clever. Prospective firebugs had a wealth of library material from which to draw. *The Anarchist Cookbook* described how to torch a building without leaving evidence, how to build bombs and incendiary devices. Never mind that some of them exploded in the bomber's hands.

Some of the other tips on starting clandestine fires actually worked, however. Masking tape, for example, made excellent time fuses which left little or no trace of its burning. So did sheets of fabric softener. One of the most clever arsonists Irik had encountered in a long time was the owner of a hotel that had gone belly up but could not be razed without a zone ruling by the city fathers. Impatient and frustrated, the hotel owner hired two men to take care of matters for him. Irik knew it was arson, but he couldn't figure out how it was accomplished.

Neighbors recalled seeing nothing unusual. In fact, the only activity noticed around the hotel occurred the day of the fire, when two men drove up in a cleaning van. They wore blue maintenance-man uniforms and carried mops and buckets. They mopped down the floors and then left an hour or so before a passing pedestrian noticed smoke oozing out from underneath the front door.

Irik finally figured it out, but could never prove it. The "maintenance men" mopped down the bottom floor with a flammable liquid and set a delayed fuse to ignite an hour after they left. The evidence went up in flames with the abandoned hotel.

Irik's was an interesting job, challenging, but laughter from the firehouse as he strolled through on his way to his office gave him a case of nostalgia for the "old days." In the ready room stocky Jim Barrett had his shaved head bent over a stack of mail-order catalogues through which he calmly thumbed. He was trying to locate the perfect Christmas gift for his wife. Irik remembered that the firefighter went through this same ritual every year. Each season he started shopping a little earlier than the previous year, but buying later. Last year it was Christmas Eve before he finally made a decision.

Barrett looked up at Irik. "What do you buy a woman who has everything?" he asked.

"Me," Merker interjected.

"We already got one of you tied up in the backyard."

Rod Harris, the shift officer, cast his fierce gaze on the fire investigator and offered coffee. Irik accepted. He liked to hang around the kitchen. Harris was one of the old-timers from the old days—a middle-aged man with a deep, intense voice improbable in such an average body. It was his unstudied opinion that all firefighters started off as arsonists.

"You still trying to catch your hotel pyromaniac?" he asked.

There had been a rash of recent hotel arsons. Fortunately, whoever started them wasn't notably sophisticated. Mostly he piled up old newspapers or cardboard boxes in linen closets or other areas readily accessible to the public and set them afire with matches. Still, Irik feared one of the fires would get out of control sooner or later and the fire department would have another Fontana Hotel fire on its hands.

Trying to catch the arsonist, Irik now rolled on every hotel alarm. He stood in the crowds of onlookers and watched faces, searching for that telltale glaze of the eyes, that excited masturbatorylike expression pyros always seemed to wear in the presence of fire. Too bad all of them couldn't be as accommodating as the naked guy on the fire engine.

"He'll show up sooner or later," Irik said, as if convincing himself.

Harris shook his head. "Maybe you'd better look

over this bunch of smoke eaters," he said. "Just pick one of 'em, any of 'em. Did I tell you about the poll I took? Ninety percent of all firefighters started fires when they were kids. That's scientific. I'm not talking about burning matches and smelling 'em either. Look at him over there."

He pointed.

"That fire-eating hero damned near burned down his own house when he was ten years old."

He pointed again.

"That one. He set the vacant lot next to his house on fire. I guess he thought he was a volunteer fireman getting paid by the job. Or like that colleague of yours in California, Irik, who got nabbed for setting the same fires he investigated."

Harris worked his way around the room, pointing and proving through rapid commentary that he had indeed conducted some kind of poll. The firefighters joked it off. Irik found it difficult to joke about arson. He kept thinking of the Fontana and the charred corpses.

Irik finished his coffee and rose to leave.

"Keep at it, Jim," he encouraged Barrett. "You'll find the perfect gift."

"And you'll find your pyro."

12

IT WAS JUST AFTER DAWN. CHRISTMAS STRING LIGHTS, chasers, adorning a small pine in a yard down the block flickered dead, going off. The sun's strip of red widened above the distant blue-green of the Atlantic. Ocean breezes rustled palm fronds in the narrow front yard fronting one of the tract homes built during the 1950s Baby Boom when the American Dream meant a patch of grass to cultivate, TV dinners, Tupperware, and 2.2 happy children. The house hunched down into its shoulders against the breeze. It was low and piss-yellow ugly, with bars on the windows.

It had become a fortress in the morning light—its doors bolted, windows closed, curtains drawn. Black-and-white police cars surrounded it. Heavily armed Miami Beach police officers crouched behind cars and hedges and blocked off the street to all traffic. Down the block a SWAT command van set up communications and issued special weapons and equipment to cops wearing dark blue jumpsuits. Neighbors watched cautiously from cracked doors and partly drawn window shades.

Lieutenant Ed Delfaverro of the Miami Beach Fire Department checked his aid bag to make sure it was ready for use. Just in case. The gunman inside the house had been holed up for over an hour now while a cop with a megaphone worked him over, begging him to give himself up.

"Mr. Morrow, we can work out the problem between you and your wife. But first you have to put down your gun and come on out."

So far there had been no response. It was too early in the morning for someone to die. It was a lovely morning.

Delfaverro stretched out his long legs on the pavement and leaned back against the front wheel of the patrol car parked at the curb. He was a big, powerful man in his mid-thirties, with a square jaw chiseled into the stony Mount Rushmore expression of a Thomas Jefferson. Calm and serious amidst the excitement surrounding him, he watched the sun rise out of the sea, above the palm fronds and the roofs of the little Beach neighborhood of tropical bungalows.

He wouldn't be needed until, if and when, gunfire erupted.

The activity had started just before daybreak when the gunman's wife scuttled out of the house shrieking an alarm that aroused the neighborhood. "He's got a gun—he's going to kill me! Help me! *Help me!*" she bellowed. "He's got a gun!"

Didn't everyone in the Miami area?

"I'll kill you bastards!" the man roared from inside his house. "Go away. Leave me alone. I'll kill myself."

Since then, the piss-yellow home sat hunched in silence while SWAT members in helmets and bulletproof vests surrounded it. The wife huddled withdrawn in the backseat of a police black-and-white parked out of sight of the house. She resembled an Okie refugee from the Dust Bowl—a hank of strawlike hair, a swath of faded cloth. She sat in the patrol car with her head hung and her face puckered like the mouth of a tobacco sack drawn closed.

That drawn face, that unhappy face, represented for cops and men like Delfaverro the dark side of the fun-and-sun capital of the world. Tourists saw the glamour and glitz of Miami Beach. Unless they were mugged or fell victim to some elaborate con game, they rarely saw that side of it away from the tourist show on Collins Avenue. Miami Beach underlife created cynicism, even among native civilians who did not have to deal with it every day. They displayed bumper stickers and wore T-shirts with cutting slogans, such as SUPER BOWL IN MIAMI—IT'S A RIOT and MIAMI: FLEE IT LIKE A NATIVE.

Delfaverro shifted his .38 caliber pistol to a more comfortable position. He felt it a forbidding sign of the times that a firefighter and a paramedic whose creed was to save lives and property also had to go armed. Fire Rescue carried bullet-proof vests, but SWAT paramedics had found it necessary to go further than that and actually strap on firearms.

Five years ago Miami Beach selected a handful of firefighters to attend the police Special Weapons and Tactics course, after which they then worked as paramedics assigned to police SWATs on emergency call-outs. The tall EMT found some difficulty reconciling his role as fire paramedic with that of an armed SWAT member.

The three-week course had been physically demanding: running miles in formation, rope climbing, rappeling, assaulting buildings, room-to-room searches, the steady drilling to mold the fire paramedics into a functioning arm of the police teams. The physical, however, had been nothing for Delfaverro. Firefighters prided themselves on staying in good condition. Tall and athletic to begin with,

Delfaverro maintained himself with regular workouts. While the police trainees used their feet in the rope climb, the firefighters used only their arms.

It was the weapons and combat training that disturbed Delfaverro. Hours on the range made the firefighters as skilled as the policemen in everything from 9mm semiautomatics to submachine guns. During the sniping phase he peered unblinking through scopes at man-targets and wondered if he could ever look into a man's eyes and squeeze the trigger, no matter how bad and evil that man might be. Just bead in on him and blow his ass into eternity.

"I'm a firefighter, a paramedic," he exclaimed. "I save lives, not take them."

Instructors set up every imaginable street situation and asked the teams to solve the problems associated with bringing it to a successful conclusion. Students practiced surveillance and hostage negotiations; they worked on hostage and kidnap victim rescue techniques; they learned how to intercept suspects and vehicles and apprehend perpetrators both with deadly force and short of the use of deadly force.

The firefighters were paramedics whose primary function with SWAT was to accompany operations and immediately treat wounds and injuries. "But you will also be going into harm's way with the police officers," instructors stressed. "What happens if you're caught in a situation in which you're the only one who can act to save an innocent person's life? What if you have to kill some asshole to save a citizen? You're going to have to be able to do that. You already know how to save lives. Now you have to learn how to take lives if it becomes absolutely necessary."

Since then the paramedics had grown more comfortable with the loaded weapons on their sides. In 1968 Miami had experienced its first racial rioting, which left five dead and nearly two hundred wounded or injured. Two or three riots in the years following the first one proved even more destructive. Civil unrest had become so common in South Florida that cops expected at least one riot a year, a kind of Orange Bowl. Cops joked about the Miami Invitational Riots.

Delfaverro had gone through the last riot and had actually drawn his weapon, he had been that afraid. That was one time he did not regret his ability with a weapon.

He still shuddered at the thought that he might have dropped the hammer on a man.

The bullhorn interrupted Delfaverro's reveries. The SWAT team was becoming restless, wanting to get this thing over with.

"What're we fucking around for?" demanded a burly cop named Doug. He had a round, flushed face, as though his helmet were too tight. "Let's go in and get that motherfucker and drag his ass out."

His partner rolled his eyes and gave a mock sigh. "Uh-oh. Doug's getting fed up. He's afraid he's going to miss his doughnuts."

In an aside, the policeman stage-whispered to Delfaverro: "Doug's always wanting to storm by force. We keep him around for balance."

"We haven't heard a word from that house in an hour," Doug complained. "The bastard has probably cut his throat by now or hung himself or something."

The other policeman knew something about suicides. "He's still in there with his shotgun. Men blow their brains out. Women cut their throats or wrists. That's a rule."

Doug snorted. "I could use a cup of coffee. And a doughnut."

Lieutenant Ed Delfaverro waited, hiding behind the police car while the sun rose slowly out of the Atlantic above the bungalow roofs. He hated domestics. Most SWAT call-outs came as a result of domestic violence —some husband beating hell out of his wife, or the wife stabbing her husband with a butcher knife or scissors. Shootings, knifings, clubbings, boiling water in the face, gasoline poured over the offending partner and a match lit . . . It was twentieth-century urban America's answer to working out family problems. Argue, hell! Go for the heavy artillery.

It always astounded Delfaverro how people who professed to once love each other could turn so savage. He thought back to the domestics he had rallied on during his past five years as a SWAT paramedic. Most of them were over by the time he arrived, but some of them returned in his nightmares.

He waited for the guy inside the piss-yellow house to either come out or turn into another of those nightmares.

13

THE DESPERATE, THE LONELY, THE DRIVEN, THE wretched. Ed Delfaverro hadn't noticed them so much when he worked fire suppression. Straight firefighters were always the heroes, the good guys. When you crept over into the police world, you entered a world that wasn't nearly so clear-cut black and white. Firefighters—just plain firefighters—were probably the last real knights in the world.

"That guy is not going to come out of there," said the burly SWAT policeman Doug.

A police lieutenant scurried stooped over from car to car, as though under fire. "Let's get ready to do it," he coached. "Let's get into position. We're going to bring him out."

Delfaverro peeped around the bumper of the patrol car. He watched police officers in helmets and black SWAT uniforms take up movement. It was swift and clean and almost a joy to watch, like choreographed ballet, this dance of impending death. Already sniper teams were in position, ready to bead in on the barricaded suspect and drop him with a well-aimed one shot-one kill. Security teams had blocked off all routes of escape and barricaded the entire neighborhood against traffic. Immediate neighbors had been evacuated.

Now, assault teams moved close with their tear gas and masks and semiautomatic rifles and shotguns. Leaders at the mobile command post had settled on a plan of attack. Delfaverro didn't need to know what it was, although he understood it as he watched SWAT cops cover various angles of the house. It would be tear gas first, followed by a charge through the front door. Classic stuff.

It might be a bloodbath in there, depending upon how determined the gunman's resistance. Delfaverro checked his aid bag again, making sure IVs were assembled and ready, the ambu bags available. He moved a stack of sealed pressure bandages to the top, where he could reach them quickly. Fire Rescue waited down the street, ready to transport casualties.

"What crimes has the guy committed other than scaring hell out of his wife?" Delfaverro demanded when he first came onto the scene. "So he's a little crazy and maybe he has a shotgun in there. That's not against the law. Why take a chance of getting people killed? Why not just let him stay in his house until he gets tired and comes out on his own?"

"Lawsuits," a cop explained.

"What?"

"The city is liable if we leave and the guy comes out and shoots somebody. The only crime he's committed so far is a misdemeanor, if that, and you know his wife's not going to file on him. But we have to arrest him anyhow—and maybe even kill him—because of lawyers. It's a Catch-22. Damned if we do, damned if we don't."

It was now full daylight. Delfaverro heard the morning traffic rush on the nearby causeway to Miami. The megaphone cop blared final pleas for the guy to surrender.

"Five minutes!" he bleated. *"Mr. Morrow, you have five minutes to put down your weapon and come on out. You will not be harmed."*

The countdown began. Assault team members crouched, tense, ready for combat.

"Three minutes, Mr. Morrow..."

Someone detected movement: "He's at the door!"

A hush fell. The surrounding gantlet of policemen bristled with ready steel as the front door of the piss-yellow house slowly opened. The cop with the loudspeaker cast it aside and aimed his pistol across the hood of his car. Nothing moved except the heavyset man who stepped cautiously onto his own porch, blinking in the morning sunlight as though he had just awakened. His gut was too big for his shirt by a hefty margin that showed between where shirt ended and trousers took up. A shotgun hung from the end of his right arm.

He looked around for a minute. He appeared confused.

"Put the gun down real slow," the SWAT commander ordered from concealment. "Go ahead, just lay it down. Don't make any sudden moves."

The fat man waddled to the edge of the porch, peering around like a myopic bird.

"Put the gun down!"

Put it down, Delfaverro begged in his thoughts. Please put it down. You don't want to die.

Abruptly, without warning, without uttering a word, the gunman spun sideways in a movement surprisingly agile for so big a man. The shotgun leaped to his shoulder. He thrust the muzzle toward the nearest policeman hiding at the corner of the house. That was as far as he got.

He must have known he didn't stand a chance. Apparently he intended to commit suicide by making the cops shoot him.

Some police departments might have opened fire en masse, giving him his wish in spades. The Miami Beach SWAT team proved well-disciplined. Only a single shot cracked. It cracked once, filling the air with the bullet's passage. Mr. Morrow's right leg crumpled from underneath him like the Invisible Man had snuck up and smashed him behind the knee with a Louisville Slugger. The jolt of the police bullet shattering bone and flesh spilled the man onto the front lawn. A chunk of his kneecap flew across the porch. His shotgun landed twenty feet away from where he fell, writhing and screaming in the grass.

A Fire SWAT paramedic was a bit like being a

military combat medic. Delfaverro jumped up and rushed forward with his IVs and bandages, the .38 revolver banging on his hip.

Somewhere in the background he heard the Okie refugee woman screaming at the police for shooting her husband.

14

DECEMBER ON MIAMI BEACH WAS LIKE A FAT MAN IN A small suit.

December on Miami Beach turned into the modern firefighter's nightmare. It was the season for fire and mayhem. The Silly Season. Christmas and disaster had a long history of running hand in hand. The two greatest single fire disasters in American history—the Iroquois Theatre Fire of 1903, in which 602 people perished, and the Brooklyn Theatre Fire of 1876, killing nearly 300—both occurred during the yule season. Probies—probationary firefighters—learned about them in fire school.

Captain Jim Barrett of the Miami Beach Fire Department had studied them when he was a probie. But *those* fires were the furthest thing from his mind when the firehouse alarm jarred him awake at four A.M. The lights came on. Men bolted from their bunks and fumbled for their bumber boots and turnouts.

Their heavy coats hung on the wall like a row of empty scarecrows.

Firefighters received little sleep for a solid month over the Christmas holidays. Fire companies and Fire Rescue squads stationed in the four strategically located houses on Miami Beach responded to thirty calls or more during each twenty-four-hour shift. A shift ran from seven A.M. one morning until seven A.M. the next.

"I'm getting real good at sleeping on my feet," firefighter Carlton Davidson decided. "I don't even have to close my eyes."

The call was for Rescue. Some female trapped in her car. Probably drunk. Barrett heaved a sigh of relief. He was fire suppression. Down the dormitory, after the three crewmen of Rescue 2 had grabbed their gear and ran, the lights went out again. Mogen groaned expansively as he climbed back into his bunk, obviously comforted that the call wasn't for his team. Someone passed gas, loudly. The house was all men, except for the blond Glenda Guise, but she had learned to live with the men and their good-humored male crudity.

"There's a kiss for you, Barrett," Daugherty quipped of the fart.

"Smells too much like your breath," Merker responded sleepily. The men muttered and grumbled—and a moment later they were snoring again. Getting in bunk time before the next call. Engaging the bunk monster.

Barrett got up and went to the window to watch the square red Rescue van leave the firehouse. Even after all this time, alarm bells and sirens still caused the adrenaline to pump.

"How did you feel on your first fire?" someone, a civilian, had asked him once.

"Scared."
"And on your last fire?"
"Scared."
"Then why do you stay with it?"
"If I knew, I probably wouldn't still be here."

Maybe *scared* wasn't exactly the right word. *Excited* was more like it. But there had been times when he was so goddamned scared that it took an hour after things were over for his heart to stop pounding.

Through the darkened window Barrett watched the Rescue van squall off the apron of Station 2 onto Pine Tree Drive. Emergency lights flashing, it blipped down through the tunnel of great Florida pines that lined the street. The pines glowed with colored Christmas lights. Palms in the yards of residences stood out against windows lighted by other colored lights or menorah candles. The lights reflected on the firefighter's broad face, drew out the dark thickness of the mustache and contrasted it to the slick baldness of a head that he shaved as closely as he shaved his jowls.

While he was shorter than the average man, he was also broader than the average man. He was built hard and muscular. He looked as powerful as a tank. No one would think such a man could be scared of Satan himself.

Barrett walked the dark firehouse. He sometimes did that while the others slept. The house had been his home—as much home as his *real* home—for the past fifteen years. He seemed the quintessential devil-may-care smoke eater who had earned virtually every certification available to a fireman. Big Jim Barrett, like that other Station 2 legend, Jim Reilly, was always first into the smoke, last out.

The deli-grocery fire had seen him first in—and almost *never* out.

Barrett walked through the quiet time, listening to

the snoring of the other firefighters. In the ready room someone had put a Santa's beard on the Hog's Head Chief. The Hog's Head Chief was a mounted boar's head that a firefighter named Charlie Brown found years ago in a tree next to the ocean on Halloween. It became Station 2's unofficial mascot. Most of the time it wore a fire chief's helmet and a yellow fire coat.

"God looks out for firefighters," Barrett always said. He smiled as he tapped a cigarette from its pack. God—and the Hog's Head Chief.

He lighted the cigarette. He gazed into the tiny plume of flame rising from the lighter. For some reason his mind's eye looked back to the deadly flames of the deli inferno.

15

FLAME GLOWED THROUGH THE BROAD FRONT WINDOWS OF the storefront deli-grocery on Washington Avenue, seeping heavy smoke into the three-A.M. sky. Station 2 Engine Company—Lieutenant Jim Barrett's company—took second out on the two-alarmer, behind Station 3's ladder company and the engine pumper from House 4.

The involved building was a three-story structure that took up most of the block. It was so old, a vacationing President Hoover might have shopped there. Smoke, thick, boiling, trying to vent, filled the lower store. Heavy plate-glass windows seemed to bulge from the heat pressure built up inside.

"That sum'bitch is really charged!" somebody shouted as Barrett, driving, braked his pumper at the curb in front of the building.

As commander of his company, Big Jim Barrett always felt a little like a military leader in combat. In fact, attacking a fire *was* a lot like a military action. The hierarchical structure of a fire department was borrowed from the military, while many of the terms —company, attack, front, assault, troops, advance, retreat, operations, tactics, strategies—might be found in any military field manual.

The major difference in fighting a human foe and a fire was that the military could choose its battlefield; in firefighting, the enemy chose the place, the time, and the conditions—and he struck always by surprise ambush. And when he struck, there was no time for second thought or casual planning. Firefighters must do something—*right now!*

Although there was little chance of anyone being trapped inside the deli-grocery at this time of morning, Station 3's ladder truck was on the scene. The ladder truck's primary responsibility was the saving of lives. After that, truckies ventilated to keep down backdrafts and flashovers and forced entries for the purpose of rescue or attack approaches.

Barrett's Engine 2 firefighters jumped off their truck at the front of the deli and began laying hose for a frontal assault. Truckies helped them. Engine 4 took the rear exposure. The danger of a two-front attack was that one company might drive heat and smoke toward the other.

Each pumper carried a recommended 1500 feet of either 2½-inch or 3-inch double-jacketed, rubber-lined hose in a double bed so that two hoses could be dropped simultaneously. A pumper also carried 1½-inch and 1-inch hose. The smaller lines were generally used for interior attacks, while the larger were used to cover outside exposures or for the interiors of large buildings.

Barrett's firefighters found the operating stem on the nearest hydrant broken. Barrett, on the nozzle of the 2½-incher, backed up by SCBA-masked Robert Plane and Mike Brady, yelled for water. Get that goddamned water started.

There was no red visible, but the charged building was clotted with smoke. It was churning in there, building up heat pressure, starving for oxygen. Heat radiated from it. It took a building fire only five minutes to heat up to 1000° Fahrenheit.

One thousand degrees!

In ten minutes it was 1300°, well above spontaneous combustion or ignition temperature for normal room contents. The building had to be cooled down quickly with a bath. Cool it down or risk flashover. A flashover occurred whenever fresh oxygen was suddenly introduced to a smoldering fire and everything around it suddenly ignited spontaneously.

"Get us some goddamned ammo!" Barrett shouted at three or four firemen, who grabbed the connecting end of the 2½ and dragged it to another hydrant down the street. Each fifty feet of the main hose weighed ninety pounds dry. Engine 4 firefighters had already launched their attack from the alley in the rear. From the front Barrett heard Engine 4's troops shouting, the forced crashing of water.

A truckie from the ladder company trotted across the street to give the deli a quick evaluation to determine ventilation needs. He peeked through the smoke-fogged window. Apparently he didn't like what he saw. His SCBA concealed facial expressions, but his body language shouted surprise. He took off like someone had fired a starter pistol.

Racing, he reached the middle of the four-lane street before the space behind him filled with a brilliant fireball. The storefront blew out, hurling glass through the air like shrapnel from a howitzer round. The force of the explosion lifted the truckie off the pavement like a rag doll and hurled him the rest of the way across the street. The SCBA face mask saved his wife from kissing a face peeled down to the bone when he struck the pavement. Nevertheless, the blow of landing was like taking a shot from ex-heavyweight champ Mike Tyson. Stunned, the truckie nonetheless managed to jump up and run.

The building lit up. Dragging the limp hose, Barrett's crew retreated before the sudden crackling appearance of flames. Shards of glass sliced out of the air around them. The heat felt like someone had opened a giant furnace. Engine 4 in the alley drove the flashover flames toward the front.

"Water! Goddamnit! Water!" Brady yelled. The lenses of his face mask reflected the fierce red of the flames.

The building really started to cook.

Barrett felt it then. The big hose filled with water, jerking him forward with its contained power. He was armed. When he opened the nozzle, the physics of water pressurized through the engine's mighty centrifugal pump, forced through a 1½-inch nozzle from out of a 2½-inch hose, produced incredible force. It could kill a man hit in the wrong place. It could tear down walls.

Controlling its working end sapped strength like wrestling with a sea serpant. Although firefighters took to nozzle time like boys tossing leftover coffee on a campfire, they had to relieve each other on the nozzle every few minutes.

Fire roared out of the deli as the heavy bolt of water ripped into its churning guts. A solid wall of bright flame snapped at the firefighters, hurled smoke as though to choke those it could not reach.

"We're going in! Get ready!" Barrett yelled above the crackling flames. His line would launch the main attack, while the second line on Engine 2 and Engine 4's two lines in the alley covered exposures.

On Miami Beach, firefighters considered themselves front-line blitz troopers. They took great pride in the fact that it was their department's policy to launch an internal attack whenever practicable.

"Give us the right equipment and we'll go right into hell and kick Satan's ass," firefighters boasted. The firefighter's world was macho, prideful. Who else threw himself into a flaming inferno when everyone else of good common sense was trying to get out?

There was no need for truckies to hammer an access to the building; the fire had done that itself. Truckies joined hose crews or shook out salvage covers in preparation for following the pipes in and saving what contents they could.

"Barrett, got your will made out?" Bob Plane asked lightly. "Hope you got your insurance made out to me."

They joked about it, about death in the line of duty, but every firefighter knew the big stats on their profession—four out of every ten of them injured each year, nearly two hundred a year KIA. Killed in Action. Another military term. Each time they entered a fire, they tempted fate and the stats.

Barrett thought he heard the dry, hollow laughter of Fate as, pipe charged, he and his crew advanced behind their stream of water toward the flames.

16

KNOCKING FLAMES BACK FROM THE ENTRANCE, BARRETT'S tiny band of smoke eaters fought their way inside. It was still a furnace, but perhaps a *cooler* furnace. Heavy smoke banked down from loft offices above the main floor. That meant there was no venting on the roof; the smoke was backing up, filling the store with an acrid and lethal fog. Barrett breathed deeply through his air hose. A tank contained about thirty minutes of air.

The seat of the fire appeared located upstairs in the offices. Flashover following the pressure explosion had ignited fire in the grocery aisles. Patches of red dripped down walls and flashed brightly here and there, especially among the cereals and paper products. Outside, truckies got the floodlights going, but the smoke reflected back the light, leaving only an eerie Dracula twilight inside the building. Barrett guided himself through the darkness and the smoke using the patches of flame. Charred debris falling from the ceiling rattled on his helmet as he hosed down the red.

The pressurized water slung canned goods; the hose caught on displays and brought them crashing to the floor. Unable to see more than three feet ahead, Barrett, with Plane and Brady on the hose behind him, followed the dim glow of fire to the hazy outline of a stairwell. Here, walls blocked the floodlights from outside, visibility dropping to even less than three feet. But there was a dull wide glow emanating from upstairs—the enemy's lair.

The second hose covered the downstairs exposure, keeping the way clear. Barrett and his men groped their way up the stairs, feeling with their boots, moving toward contact. Smoke closed in around them, isolated them. Even experienced firefighters had died after becoming disoriented and confused in smoke. There was one story about a fireman who became separated from his hose and died trying to find his way out of a walk-in *closet*.

"The hose is your lifeline," fire instructors preached. "Stay with it. In smoke where you can't even see your own hand, you can at least follow the line back out to safety."

The three men dragged the bloated hose through smoldering ruins. Barrett knocked out extension flames that had clawed their way off the main fire, eating into walls and floors and ceilings. The wooden stairs felt relatively sound. Upstairs, dimly, orange glowed through an office window down a narrow hallway where smoke hung as thick as heavy drapery. The office across from it also crackled with fire.

Barrett opened the nozzle wide. He heard window glass shattering. The fire hissed at them, turning water into steam. Water rivuleted around their feet and cataracted down the stairwell, as though in fearful flight. The men fought with heavy bursts, fighting and inching their way deeper into the dark hallway toward flames raging inside the two offices.

Barrett switched places with Brady to let Brady get in some nozzle time. The floor shuddered underneath Barrett's feet, knocking him against Bob Plane. The three firefighters grabbed fistfuls of each other's turnout coats for support as the floor shuddered again. Then they heard a deep rumble that seemed to start somewhere in the grocery's bowels and work its way out until the entire structure shook like an old man about to collapse with congestive heart failure.

The firemen exchanged quick anxious glances. The whites of their eyes showed through the smoke-fogged lenses of their SCBAs.

"The roof—it's coming down," Barrett said tersely. "Let's get the hell out of here."

"Feets, don't fail me now," Brady retorted.

As experienced firefighters, they realized immediately and with sudden dread what had happened. The fire had lured them into a trap. The fire you could not see was often the most dangerous. It spread very rapidly through pipe shafts, ventilation ducts, conveyor openings, between partitions, and through ceilings and roofs. It ate out the bones of a building so that by the time it was noticed, the building was ready to collapse.

This building was about to go.

Before the firemen reached the return head of the stairwell, stumbling in the dense smoke, running their hands along the hose for guidance, they were shook nearly off their feet by a second rumbling. Something that sounded a bit like a Cadillac being hurled through the ceiling landed with a terrific crash in the store aisles below. Debris fell like hailstones around the firefighters. They ducked instinctively, protecting their faces, but there was nowhere to hide.

Fire crackled behind them with renewed energy.

Ahead, down the stairway, lay a black curtain of smoke. Part of the roof had caved in. It was only a matter of time—minutes? seconds?—before the rest of the roof joined it.

Barrett led the way, feeling his way on the stairs. The hose in his hands had gone limp after the cave-in, obviously pinched off or cut by the collapsing roof. It felt limp like a dead python. Without water they were defenseless against the flames. The faint glow through the smoke seemed to pursue them.

"About your insurance, Jim . . . ?" Plane muttered, but his voice was tight, without humor.

So far, the stairwell remained intact. The men knew it led into the canned good aisles; only thirty feet from the windows and escape. Following the hose, running his gloved hand along the wall, stocky Barrett reached the bottom of the stairs—and crashed headlong into a solid barrier.

"Christ!"

He bounced back against his companions. They went down in a tangle of arms and legs, from which Barrett quickly extricated himself. His gloved hands reached out in total darkness, exploring the barrier. His heart sank as he reached as high as he could and to either side and found nothing but a solid wall of debris. He kicked at the wall in frustration. It held firm.

"The goddamned roof, ceiling, everything has come down," he exclaimed, trying to remain calm.

He didn't say it, but the thought clamored in his mind, blinking on and off like a neon sign: *Trapped!*

Trapped between the blaze upstairs and the fallen building between them and the exit. As if that wasn't enough, adrenaline rush caused him to suck air like a racehorse at Hialeah blowing after a run. With a feeling that sank all the way to the pit of his stomach, he tasted the bad air left at the bottom of his air tank. Murphy's Law. Anything that could go wrong—did.

He calculated he had about five minutes of breathing time left. Nine out of ten injuries firefighters suffered was due to smoke inhalation. Fire consumed oxygen at an alarming rate, replacing breathing air with toxic gases like carbon monoxide, hydrogen sulfide, sulfur dioxide, and other gases, depending upon what was burning.

Five minutes of life remaining. The firefighters' masks clacked together so they could see each other's eyes.

"How's your air?" Barrett asked the others.

"It's already tasting bad," Brady responded. His eyes flicked. Plane nodded.

"Mine too. Okay, we can't stay here like rats."

Barrett struggled to keep the desperation out of his voice. He had seen fire corpses before. He grabbed his companions and gave them a little shove. Together they traced the limp hose back up the narrow stairwell. The fire was really starting to cook again upstairs, even though Engine 4 remained busy outside pouring water onto exterior exposures. The entire building was now at risk—as were the three hapless firefighters trapped inside it.

Even though the truckies were undoubtedly attacking the problem with their axes and hooks and ladders, Barrett realized they could not depend on outside help for their rescue. They didn't have that kind of time. The second-story floor shuddered underneath their feet; flames glowed dully through the rolling smoke. Heat blasted against their bodies. In just minutes they would be forced to breathe air as toxic as that on Mars.

If the building lasted that long.

Feeling their way along the walls like blind divers, almost swooning from the built-up heat, choking and coughing from the bad air and the acidity of the smoke, which was already beginning to burn their lungs and eyes, the desperate trio tap-tapped its way to a small window that opened onto Washington Avenue below. The window was so small that a man's head would not fit through it. Even if they dropped a line for more air, they couldn't get the bottles through the window.

The face of the truckie who had been knocked off his feet when the window exploded stared anxiously up at the window. He yelled something drowned out by the fire. Men came running with ladders and axes. Other men crowded at the front of the building, hacking away at the debris, trying to get through.

Brady stumbled against Barrett. He made a knife-hand slash signal across his throat. Dizziness also welled up in Barrett, coming over him like a blanket over a child. The world swirled as if he had drunk too much.

Fire closed in on them. Even the few minutes it took truckies to scurry up ladders and chop a hole through to them would be a few minutes too long.

It would be so easy to surrender to the smoke, but a near swoon shocked Barrett with adrenaline.

Like rats . . .

The firefighters shook it off and for the second time followed their hose to the fallen ceiling below. They knew they would not be coming back upstairs, no matter what. The lower they were, the better their chances of finding good air.

Not content to simply lie down and give up, fighting for their lives no matter how hopeless it might appear, first one firefighter then another attacked the smoldering rubble with their hands until they were all digging like beagles after a gopher. They felt the heat behind them, driving them on.

Even in heavy gloves their hands immediately grew raw and scorched. They gasped for breath.

"Here!" Plane yelled out hoarsely. "Here! *Here!*"

Sudden hope. He directed Barrett's hands to an opening. Barrett reached into it as far as he could. It appeared to be a crawl space just large enough for a man to squirm through on his belly if he dropped his air pack. The only question was, did the tunnel extend all the way through to safety?

There was only one way to find out. Behind them fire dripped down the stairwell. Without more than a pause to clack masks and give each other a last look, the men dumped their air, Barrett first. Instantly, smoke gripped his throat like claws around his neck. Funny, he thought, how the situation had changed in just a few minutes. They had had the fire backed into near submission, but then the enemy struck from a new quarter and retook the battlefield.

He dropped onto his belly and followed his companions blind into the smoke-filled opening. It closed in on him, narrowed. He suffered a moment of claustrophobia when he knew, he *knew*, that the rest of the building was coming down. They would find his body and Brady's and Plane's underneath the rubble, and the chief would make his obligatory personal visit to their wives.

Then he heard someone shouting, laughing with relief. A moment later he emerged on the other side, coughing and hacking, into the arms of the truckie rescue squad.

He was alive.

God, indeed, looked out for firefighters.

17

FIREFIGHTERS HAVE ALWAYS BEEN FASCINATED WITH NEW techniques and methods of combating fire. As the bucket brigade led to the hand-operated pump and the hand-operated pump to the centrifugal pump, so has common water, still the firefighter's primary weapon, led to experimentations with chemicals and other extinguishing agents. An important part of fire history is the development of chemical engines, which helped pave the way for the use of modern specialty fire weapons.

In 1864 two Paris engineers, Dr. F. Carlier and A. Vignon, combined sulfuric acid, "oil of vitriol," with bicarbonate of soda—ordinary baking soda—to form a chemical reaction which yesterday's fireman called carbonic acid gas. It was really carbon dioxide. Placed under pressure of 200 pounds per square inch, the carbon dioxide mixture led to the "soda-acid principle" of extinguishing fires, which in some places in the United States lasted into the 1930s.

Shortly after the chemical discovery, in 1868, the New York City Fire Department had soda-acid fire extinguishers installed in every fire station, where they were placed next to the watch desk. When the alarm came in for any minor fire, firefighters grabbed the bottles and raced on foot to put out the blaze.

Two years later the Boston Fire Department began using a horse-drawn "fire-extinguisher wagon" which carried ten hand-operated soda-acid extinguishers. It was the only one ever placed into service; the next year, 1872, Babcock & Holloway of Chicago began constructing hand-drawn and horse-drawn chemical engines. The apparatus carried a large tank reservoir of the chemicals, to which spraying hoses were attached.

For more than a half century the chemical engines remained an important fixture of any "modern" fire department—until the final realization that, when it came to extinguishing fires, soda acid was no more effective than plain water.

Undaunted, however, firefighters and scientists continued to experiment. In addition to water, today's firefighter has available in his arsenal a variety of weapons to aid him in the battle against man's oldest enemy. Science has discovered ways to make water "wetter" and therefore more penetrable, thicker to give it greater power to cool and smother, and "slippery" to reduce friction loss in hose lines and increase water flow. For example, a 1¾-inch hose using "slippery" water can pour as much water on a blaze as a conventional 2½-inch hose.

Science has stocked the nation's fire arsenals with chemicals and gases that do not conduct electricity and therefore can be used on electrical fires; with foams to smother flammable liquid blazes, and with high expansion foams for underground oil fires; and with dry powders to combat fires like those involving magnesium, titanium, sodium, and potassium, which are very reactive to water and are capable of generating their own oxygen supply.

"Science may replace ordinary water," Captain Jim Reilly observed once, "but there is one thing it can never invent or replace—and that's the firefighter. You can bet your rosy ass that however long this civilization endures and that no matter how advanced we become, when the alarm sounds at the fire station it'll be live firefighters who respond to risk their lives to save your wife and kids and the family heirloom."

18

"Hey, Glenda. Are you decent?"

"You know I'm not. Come on up."

As the first and, to date, the only female firefighter admitted to the Miami Beach Fire Department, blond blue-eyed Glenda Guise occupied a special place, whether she wanted it that way or not. She had invaded an all-male domain. At first the men disconcerted her. In many ways they remained little boys who had simply grown up to ride the fire trucks they used to watch go by.

Downtime between alarms, they could only do so much cooking and housecleaning and shining and working out in the gym room. All that excess energy had to go somewhere. Part of life at the firehouse was the joking, the pranks. Short sheeting the bunks. Waterfights. All that, plus sleeping and showering and going to the toilet and working with a bunch of brawny tough guys, took, for a woman, some getting used to.

Having a woman around had taken some getting used to by the firemen too. Cops and combat soldiers and firemen were a macho breed who spoke rough and kicked ass if they had to. They were not Ph.D.'s and neat little buttoned-up men with soft white hands, pointy-toed shoes, and the leisure and inclination to intellectualize and go out of their way to be politically correct. They functioned direct-action, viscerally, and sometimes, in their vernacular, a spade was not a spade—it was a goddamned shovel.

Back when the firefighters at Station 2 first heard that a woman was being considered to attend fire school, they looked at each other, stunned, and shook their heads. In spite of lawsuits, Miami Beach had always refused to hire firefighters based on quotas for race, ethnicity, or gender. Cynical conversations broke out among the men like an epidemic of measles.

"What's she, a bull dyke? She think she's got balls like a man?"

"I hear she's a twenty-one-year-old blonde. What the hell she want to be a fireman for? Doesn't she know that's how it's spelled—F-I-R-E-M-A-N?"

"Maybe they're going to issue a blonde to each of us to keep us happy. A horny firefighter is not a happy firefighter."

The woman who applied for the firefighter position at Miami Beach City Hall that morning eight years ago possessed a wholesome girl-next-door quality that fit well with the well-built body only slightly taller than average. She had been a lifeguard in high school and played intramural sports. For as long as she could remember, even the thought of some indoor job traditionally reserved for women—secretary, nurse

—sent chills down her spine. She wanted a challenge, not a job.

Her husband Gary had enlisted with the Miami Beach Police Department. He soon switched over to the fire department. Glenda suggested that she could be a firefighter.

"There are no women on the Beach Fire Department except those who work in the offices," Gary said.

"Then it's time for a change."

She first attempted the City of Delray Beach, north of Miami. The personnel department laughed at her. Miami Beach didn't. She took the written civil service exam and passed it among the top ten percent of the other applicants. She was the only female. She breezed through the personal interview.

"I'm not asking for any special considerations just because I'm a woman," she earnestly assured the personnel board. "I'm simply asking for the opportunity to prove I can be a firefighter."

Irish Fire Captain Jim Reilly, before he died, bluntly voiced the sentiments of most firemen when he observed, "By God, you become a firefighter *here,* it's because you got the stuff it takes, not because some idiot court decides your color or sexual preference or something is underrepresented. We don't give a damn if you're a little green man or an hermaphrodite, but we'd better be able to depend on you in a fire. You'd better be able to fight fire."

Most firemen simply did not believe a woman could do that. They watched Glenda's progress through the selection process with skepticism.

"All right, so she made it through the physical agility test," they commented. "She'll never make it through fire school. Even if she does, she ain't got the balls to lug a two-and-a-half hose up a tenement stairs. Firefighting is a job for *men.*"

Until she busted into an all-male domain, Glenda failed to realize the impact a woman made. With all the publicity and fanfare attending her first week at the six-week fire school, some of her male students were jealous and angry. They didn't understand; no *real* woman would want a *man's* job.

"The city bowed to pressure," went the comments, sometimes in Glenda's presence. "She's our token to appease the women's rights movement. She won't last."

The female probie ignored the remarks, pretended not to hear them. Responding to them only fueled resentment. She knew one thing—that as a woman, and as a prospective firefighter, she could not afford to fail. It wasn't good enough to make it through fire school in the middle of the class. She had to excel. Everything in her, every reserve, had to be directed toward a single goal: making it through fire school. She had an exciting career ahead of her—if she could get that one obstacle behind her.

She would *not* fail.

"Are you trying to prove that women are equal to men when it comes to all-male jobs?" a reporter asked her.

"I am no feminist," Glenda responded. "I'm not doing this to chart some course for other women just so they can say they did it. The only point I'm trying to make is that I can be a firefighter. Don't call me a feminist; call me a firefighter."

Being a firefighter could be hard work. Training for it was even harder. Fighting fires, the young blonde discovered, involved more than simply learning how to wear fire gear and knowing the difference between a fire engine and a ladder truck. Probies studied classifications of fires according to the materials and structures that were burning; the four stages of fire, beginning with the incipient stage and ending in the final stage with a conflagration that consumed everything.

New York's Great Fire of 1835 was an example of a conflagration. It was the most destructive nonmilitary fire the world had known since the London Fire of 1666. It raged out of control for fifteen hours, leaving seven hundred stores in ashes in the First Ward. The losses led to the Wall Street panic of 1837, the most serious depression up to that time.

Curiously enough, there was only one fatal casualty. A crowd discovered a man setting a fire, whether in arson or as a backfire against the main fire, no one bothered to ask. The mob lynched him and left him hanging for the fire to consume.

Students pondered over building materials and construction and how different types of material reacted to heat. They studied fire science—how fire traveled and behaved under different situations, the behavior of smoke and flames and the many toxic gases produced. They learned which windows blew most readily under pressure, and about fire walls and fire safety procedures.

"We call this the fire triangle," instructors lectured, pointing to diagrams on the blackboard. "It takes three things to make a fire—air, heat, and fuel. The fundamental method of extinguishing flames is to remove one side of this triangle. You can take away heat by cooling the fire with water. You can smother the fire with dry chemicals, robbing it of oxygen. Or you can starve the fire by denying it fuel. Sound simple?"

It wasn't, Glenda discovered. Maybe in theory, but not in practice. The fire students spent long hours in the classrooms, learning such fundamentals as the three types of hose nozzles and the inverse relationship between gpm (gallons per minute) and psi (pounds of water pressure per square inch). When the gpm increased, the psi fell, hoses and hydrant pressure being equal, and vice versa—when the psi increased, the gpm decreased. Increasing the nozzle pressure placed a stronger stream on the fire, greater psi, but decreased the amount of water, the gpm.

Most fires, they learned, could be controlled with 30 to 100 gallons of water per minute, while a good working fire might take up to 1000 gpm. The 1934 Chicago Stockyard Fire required 50,000 gpm.

They learned different ways to attack fire. The pumper or fire engine was the first line of attack. During an assault, one main line was directed at the fire, while successive lines covered exposures—rooms, walls, and buildings adjacent to the fire itself. There was little value in extinguishing fire on one floor only to have it break out two floors up because of not covering exposures.

A one-front attack was sufficient for most fires. The first engine on a scene attacked the front of the fire, not necessarily the front of the building. Additional engines and lines covered other fronts in sequence: first, a line inside from the front of the fire; a line above the fire to cover upward exposures; a line from the rear; a line above from the rear; and then lines from the sides.

After a fire was confined, the company with the best chance of reaching the seat of the fire pressed the attack while the other companies fell into support and reserve and covered exposures.

Students worked with the hook-and-ladder trucks as well as with the pumpers. The modern hook-and-ladder carried a standard 200 feet of ground ladders, including one-piece straight-frame ladders, extension ladders, roof ladders, and usually a 100-foot hydraulic-lift ladder in three or more sections. Many trucks also came equipped with a snorkel unit—an elevating platform—which could be used either for rescue or to lift troops and their master streams over the top of a fire.

Compartments on the trucks contained additional equipment, such as extinguishers, both Class A Water and Class B Multipurpose Dry Chemical; pick poles, pry bars, axes, hooks, claw tools, and power saws with which to open floors, walls, and ceilings for ventilation or rescue; aerial pipes or turret guns; cutting torches, power wenches, SCBAs (self-contained breathing apparatus) and air tanks; lifesaving equipment such as resuscitators and inhalators, as well as overhaul and salvage equipment such as salvage covers, water vacuums, squeegees, brooms, shovels, mops.

Fire students took to the outdoors to put into practice what they learned. Since one fire is very similar to another fire, Glenda and her classmates learned that the secret of successful fire suppression lay in training and drill, plus more drill, and then some additional drill. Firefighters preplanned their attacks on different types of buildings and for different standard situations. They followed set procedural outlines, then amended those outlines to fit a particular situation.

Glenda humped hose, packed hose, and lay it. She practiced lugging an eighty-pound ladder, slamming it against the side of a wall and scurrying up it like a monkey. She used axes, SCBAs, hooks, claw tools, power wenches.

Being a firefighter required a certain amount of raw strength and endurance. Glenda was no weight lifter, but she had always worked out to keep in shape, a habit she carried with her into fire school and afterward. What she lacked in sheer muscular power, she compensated for in enthusiasm and fortitude. The men watched her working with the ladders and hoses. Occasionally at first, and then more frequently as the school continued, she noticed a grudging smile of approval from her classmates or the half-offered compliment of a nodding head.

Firefighting, she admitted to her husband Gary, was not a job for the average woman. Most women accepted for fire schools in the United States flunked out. You had to want to be a firefighter to endure the rigorous training; you had to *want* it.

"That little girl just won't give up," Glenda overheard her classmates discussing her.

She thought about it, though, giving up, when her muscles ached and it seemed she might not be accepted by the men. But she knew she wouldn't give up. She couldn't.

Like all students, Glenda's first actual taste of structural firefighting came on the day of the smokehouse. The smokehouse was a square metal building three stories tall with openings for windows and doors and a metal stairway inside. It was in the smokehouse where instructors ignited used oil or old tires and other fuels to simulate the heat and blinding smoke firefighters often faced in actual combat. In combatting these fires, using the tactics they had learned, students received their first sense of what it was like to enter a totally hostile environment.

Each student was required to enter the smoke and remove his mask. Glenda's heart pounded when it came her turn, but she charged into the thick black smoke banking around her. It was the first time claustrophobia almost caused her to panic. She gripped herself, inside where it counted, and bravely ripped off her mask.

Smoke instantly drew tears to her eyes, blinding her. She held her breath as long as she could, standing in the smoke, head bowed and one arm reaching out as though for support. Finally she had to breathe. Heat seared into her lungs. Coughing and choking, she stumbled outside into the fresh, biting air. Her face was grimed black like a coal miner's.

"Hey, Glenda. Nice makeup," another student quipped. His face was as black as hers, so that his teeth flashed when he grinned.

Glenda grinned back. She walked a little taller. Joking was the first indication that she was being accepted, at least by her classmates.

It was a different matter within the fire department.

"Okay, so she made it through fire school," the hard-liners conceded. "That doesn't mean she'll make it through probation. Wait'll we get an all-nighter in some hotel and she has to lug hose upstairs to the fifteenth floor. Just wait."

19

AFTER SHE GRADUATED FROM FIRE SCHOOL, GLENDA GUISE had carefully donned her new pressed blues and prepared to report for duty at Station 2 on Pine Tree Drive. She pulled her blond hair back severely and wore almost no makeup. She tried to hide her femininity, make it less apparent. She didn't want to let the fact that she was a woman get in the way of her acceptance. Her husband watched her. He would be housed at another station. Civil Service regulations forbade relatives serving together.

"Scared?" he asked.

"Me? Of the guillotine? Never!"

"They'll love you," he promised. "You'll be a great firefighter."

She smiled nervously. "At least I have you convinced."

Like most probies, Glenda spent the first few weeks of her probation getting oriented to the fire department and its policies and procedures. She worked with both fire suppression and Fire Rescue, observing. Later she took additional classes in medicine to further qualify her as an EMT—an emergency medical technician. On Miami Beach, firefighters rotated between fire suppression and Fire Rescue, between fighting blazes on South Beach and reviving heart attack victims in the delis and prying auto crash survivors out of mangled steel on the causeways.

The first day at Station 2 proved an awkward time. Glenda would always remember it. The male firefighters didn't know how to receive her. They seemed a friendly bunch, some of them smiling shyly. They stood with their big hands hanging, looking around everywhere except at her. They looked at the shining red apparatus and twin Rescue vans parked in the garage, at the Hog's Head Chief in the ready room, but not at each other and not at her. The entire station turned out for the occasion. In spite of the tough talk the men had indulged in about her before she arrived, Glenda had the odd feeling that they were seeking her approval as much as she sought theirs. It was like, since she was the first woman to break into the department, the firefighters felt *they* had to live up to *her* expectations. After all, their house had been chosen for the honor of the first female.

And she was so young and pretty too. Not at all like the bull dyke image some of the firefighters had concocted for her, sight unseen.

Someone explained that the Station 2 complex on Pine Tree also housed the fire department's command and administrative headquarters—records, personnel, Chief Brainaird Dorris's office, Fire Investigator Vance Irik's base. Lieutenant Rod Harris was the station company commander of the shift. He greeted the new probie with a voice so deep and commanding that Glenda all but fell back from him. Then she looked into the middle-aged man's dark intense eyes, saw humor and patience, and immediately liked this otherwise nondescript man.

He smiled a friendly smile.

"Mrs. Guise, welcome to Fire Station Two. You'll

be working here. Hopefully, you'll like it. Just do your job. That's the important thing. You do your job and they'll all come around to your side, I promise you that. In the meantime, you're going to have to have a little patience and understanding. You'll have to overlook some things—like their language and so forth. It's going to be a little inconvenient for you and for them, maybe even a bit embarrassing at times. You're not going to get any special treatment just because you're a female. Don't expect it. If you're going to be a firefighter, you've come to the best house."

"Sir, I don't expect any special treatment."

Harris looked at her. He nodded thoughtfully. "After all," he said, "you're the first lady."

For this first lady of Station 2 there were toilet arrangements to make, sleeping arrangements. Somebody placed a lock on the toilet-shower door for her benefit. She used lockers to give her some privacy. The men went out of their way to make her comfortable, even to the point of getting up to give her a chair whenever she walked into the ready room.

"Please?" she begged each in private, whenever these things happened. "I need to be treated just like everyone else."

Firemen held rather old-fashioned ideas and values when it came to the opposite sex. They would have made great knights.

For her part, though, Glenda mostly kept her mouth shut. Having a woman in the house cramped the men's style—at least at the beginning. Normal language around the firehouse could be colorful, the horseplay rough and sometimes lewd. Some of the guys liked to watch steamy films on TV. All this they curtailed at first, but as they grew accustomed to her presence over a matter of months, their behavior gradually returned to normal. Glenda smiled and accepted life at the firehouse with quiet grace and a good sense of humor.

"I think it must be harder for the guys than for me," she confided in her husband. "I keep thinking, 'Well, how'd I feel if I was a guy and they sent this girl in?'"

Women not on the job were curious. Some of the first questions they usually asked were, "How do you do it—all those guys? How do you keep them fought off? Do you ever sleep with them?"

"Every night I'm on duty," Glenda sometimes responded with a quick grin, just to see the reaction. After all, the firefighters all bunked out in the same dorm.

Glenda was both young and attractive, with a vivacious personality. Some of the men tried to make hits on her, but she fended them off in the same low-key way she repelled insults—by simply not responding to them. She let it be known in a polite but firm way that she was married and intended to remain that way, that she did not fool around. Once that was established and her place hewn in the structure, the men began treating her much the way a large family of brothers might treat a kid sister.

One evening about six months after her arrival at the station, she finished at the sink making coffee or cleaning up the kitchen—the usual probie stuff—and ambled into the TV room. Just as she walked in, one of the guys lifted his leg and ripped off gas like a ruptured line in a fire.

"There's a kiss for you, Chapman."

"Thank God it's a kiss. I thought you were rotting."

No one giggled or sniggered or cast a side glance at Glenda. She smiled secretly. To her, it was a significant sign that she was becoming just another firefighter.

She was also perceptive enough to understand that

acceptance at the firehouse was one thing, while acceptance as a full-fledged smoke eater was another. Sooner or later every probie had to pass his trial by fire, had to prove himself. When her time came, Glenda realized that as the first of her sex to invade the fire department, she dared not be simply adequate. She had to be the best there was.

She confessed her fears to her husband. "They'll all be watching me to see how I do."

Probies were usually assigned to peripheral duties at a fire—laying hose, tending hydrants, reserve standby. Finally, however, Glenda's day of trial by fire arrived when a warehouse caught fire on South Ocean Drive. Flames had pretty much consumed the entire building before it was noticed and reported. It wasn't much of a challenge, although it shot flames thirty feet into the air and mushroomed smoke that hung over the city like an A-bomb explosion. Firehouses 2 and 3 responded to the two-alarmer. Sections of the wall and the roof had already collapsed. The building was a total loss. The only action firefighters could take was to protect exposures to prevent flames from spreading into adjacent structures. Gaps of only about three feet separated the conflagration from warehouses on either side. The neighbors were already charring.

The word came down. "Surround and drown. It's a goner. Confine the fire to the one building."

It was a perfect fire in which to break in a probie. Not exceptionally dangerous, since it called for only an external attack. Ground troops attacked defensively with a 2½-inch line, while Glenda the eager probie and several smoke eaters from Station 3 volunteered to launch an aerial assault. Tower ladders—long booms with platforms at the end—lifted fighters and their lines to the flat roofs of adjacent buildings, where they dumped thousands of gallons of water down onto the boiling smoke and flames.

It was Glenda's first time in real fire battle shoulder to shoulder with her compatriots. It proved exciting, although stamina was about all the fight required. The heat was like standing inside an oven turned to Broil. Sweating profusely inside her turnouts and SCBA mask, Miami Beach's first female firefighter could not help wryly recalling how she had always disliked working inside artificially chilled buildings.

She shared nozzle time with other firefighters, retreating from the roof only long enough to change air bottles and gulp a quick Gatorade. Eager to prove herself, she volunteered to stay behind with Firehouse 3 for mop-up, salvage, and overhaul when the attack team commander released Station 2.

One main line remained on the roof, with two firefighters—Glenda and another eager private with whom the firewoman had been battling for the past two hours. Because of the thick smoke and their face masks, they hadn't bothered to recognize each other. They were too busy. They switched off on the nozzle, Glenda taking over, and together dragged the heavy hose to a better position from which to dump water onto the smoldering rubble below. Their masks clacked together.

Glenda looked up and grinned suddenly, recognizing her own husband through the lenses of the other mask. Gary grinned back.

"Thanks, partner," he said, observing, "You are one hell of a firefighter. I'll fight fires with you anytime, lady."

The compliment meant something, coming from her husband, but she knew she still hadn't proved anything to her own company. She was disappointed, when she returned to the station for a cool shower and a rest, to find that few of the other firefighters gave her so much as a second look. They were cooking dinner and cleaning the engine and their gear in preparation for the next alarm. Her adrenaline still raced.

"Pretty good fire," she said casually, fishing for acknowledgment of her skills and endurance.

"It was a pot burner," said feisty Stu Merker, who then grinned behind Glenda's back and winked at big John Creel.

20

HARD STATISTICS REVEAL THE NATURE OF THE FIREFIGHTer's enemy. Three million fires erupt each year in the United States, accounting for over four *billion* dollars in property losses and some twelve thousand victims. Hundreds of thousands are injured, one-third of whom are children. Smoking and matches are responsible for the greatest loss of life and top the list of the major causes of wild fire. Electrical fires come next on the list, followed by heating and cooking fires and then by children playing with fire.

The majority of fire calls are for nonstructural fires such as automobiles, grass, and rubbish. Probationary firefighter Glenda Guise combatted these "trash" fires with her Station 2 crew and bided her time for that day when she knew she would have to prove herself with a major flame or fall flat trying. Being the only woman, she felt she had to work harder than the men. She always volunteered for overhaul, cleanup, and salvage after a blaze. That was the dirty part of firefighting, the unglamorous side.

Even "pot burner" fires—small fires confined to the kitchen or the bedroom or whatever—required overhaul. Large fires required salvage of whatever remained of value. The buildings had to be completely ventilated of smoke to minimize smoke damage, and then all openings to the weather covered. Salvage crews drained off excessive water, even siphoning it from below-grade levels. The object was to extinguish the fire with the minimum force necessary and then return the fire-stricken building to its owner in the best possible condition.

Barrett, Chapman, Merker, and the others worked shoulder to shoulder with the female firefighter. "That girl's no shirker," they commented.

"And she's so pretty too."

Apartment and hotel fires ranked second nationally, just below private dwellings, as a major source of fires. On Miami Beach, however, so crowded with hotels and high rises that the island seemed to be sinking from its own weight, dwelling losses may have come second to hotels and apartments. When the Saxony Hotel burst into flames against a clear midday sky, calling for a basic attack team of two pumpers and a ladder truck, Glenda hooked her arm onto the hand railing of Engine 2 and rode with the fear that, being a probie and a woman, she would be assigned to exterior exposures.

She longed to try her skills on a direct, interior, frontal assault against the enemy. How else could she prove herself?

From blocks away, riding the sirening engine, she

saw smoke rising above the city, clouding against the blue sky. Firefighters learned to identify fire by its odors. A deep-seated wood smell generally meant a big fire. Glenda sniffed the big fire smell and her heart pounded. She glanced up ahead at Barrett, who was driving. His shaved head gleamed. He said something to Merker seated next to him. They both grinned.

Glenda wondered if she would ever stop being nervous en route to a fire.

Police were busy throwing up barricades to catch and hold a rapidly growing throng of onlookers. Engine 3 was already on the scene; firefighters scurried about hooking up lines and studying the building for ventilation.

The Saxony, backdropped by the Atlantic Ocean, took up most of the block. Its great area diminished its height and made it appear square and squat. Smoke pumped thick from the open lobby doors, while more smoke bent itself out of several different windows on floors higher up. Such multiple fires often signal the work of an arsonist. Vance Irik of the arson squad had already been summoned.

Fortunately, the fire had not occurred at night, when the rooms were filled with sleepy tourists. The manager, gesticulating wildly, yelled at Engine 3's officer, telling him that he had emptied the hotel of guests and employees as soon as the first whiff of smoke set off the detector alarms.

"We checked all the floors," he shouted with self-important bravado, "and I'm sure there ain't nobody left inside."

No flames showed their heads at the windows as the pumpers and the arriving ladder truck prepared for the attack. There was just the smoke. Truckies dogged ladders and raised other truckies in the tower ladder to ventilate the roof. Without ventilation, fire simply smoldered in waiting, charging up the sealed rooms with tremendous pressure. It waited, curled up like an awakening and hungry bear in its cave, until oxygen was introduced like a bear's first meal after hibernation. Then things went to hell, literally. A room simply flashed, igniting in a sudden deadly furnace that simultaneously consumed everything within reach. A flashover fire, a backdraft when fire suddenly exploded toward fresh air, was the firefighter's most feared nemesis.

There were other dangers as well in the unseen fire. Driven to concealed chases in their search for oxygen and fuel, tendrils of silent flame gnawed into the inner spaces between floors, walls, and ceilings. It undermined like termites, until nothing remained of structure except a thin shell ready to give in and crash through with the first weight.

As if flashovers and backdrafts and weak floors weren't enough, the firefighter who entered a flaming building for an internal assault also faced smoke, which accounted for nine out of every ten firefighter injuries, poisonous gases, collapsing roofs, explosions, dehydration, and heat stroke. Entering a fire was not a natural thing; it went against all survival instincts. Beasts went berserk fleeing it; some people also stampeded, trampling anything and anyone in their paths.

In 1903 more than 1700 people were seated in Chicago's Iroquois Theater and another 200 were standing in the aisles waiting to see a play called *Mr. Bluebeard,* starring the leading vaudevillian of the day, Eddy Foy, when an arc light backstage burned its fuse. Sparks ignited hanging gauze. The fire spread quickly. Within twenty minutes 602 people were killed—from smoke inhalation and from being trampled to death. People were piled up three and four deep in front of exit doors that opened inward instead of outward. Others suffocated after massing into narrow hallways. Some never even left their seats before succumbing to heat and smoke.

Only the firefighter went into a fire when everyone else was trying to get out. Probationary firefighter Glenda Guise wanted to enter a big blaze and fight it *mano y mano* more than she had wanted nearly anything else in her young life. She was like any other soldier wanting to prove herself in the only way possible—by direct and personal combat.

"Guise, Chapman, get ready. Your crew is first in."

Glenda blinked at Lieutenant Barrett, shift officer for Engine 2. She grabbed her SCBA and slipped her sweating face into it. Her lungs sucked the bottled air. Smoke dark and forbidding reflected itself in the lenses of her mask.

Through the breach, she thought . . . at last.

21

ENGINE 3 ATTACKED THE FIRE'S REAR; THE LADDER SNORKEL shot a master stream from the air against upward exposures; one of Engine 2's lines assaulted the fire's front. Chapman and Guise and a third firefighter wrestled the second 2½-incher toward the lobby doors, laying down a stream of cover water as they advanced. Windows popping out on upward floors sounded like gunfire and enhanced the impression of full-scale battle.

There was no talking. This was where training came in, and the many hours of practice drill. As the solid wall of smoke roiled out and around her, enveloped her and drew her stumbling into its depths, Glenda experienced a moment of panic. She felt an almost uncontrollable urge to turn and flee, as she had fled the tin smoke building during training when she removed her mask.

Her breathing rasped through her SCBA. Smoke was so thick she felt like a bat with malfunctioning radar. She actually heard her heart pounding, louder than the crackling sounds of the fire still hiding in the smoke, waiting to counterattack.

The moment of panic passed. In the lobby there was just heavy smoke and heat. While her heavy turnout coat shedded water and protected somewhat against heat, embers, and sparks, it provided virtually no protection against extreme heat or flame. She felt encased in a steam sauna.

She clung to the hose behind Chapman on the nozzle and felt power throbbing in the line as Chapman opened it wide. Gradually, training took over and she relaxed a bit. A firefighter never truly relaxed in a fire, it was unnatural, but she accepted it and pulled her own weight as, with the first-in assault, she progressed into what appeared to be a narrow hallway. In the blinding smoke, they stumbled into doors —lockers or cabinets or something—that lined either side of the passageway. Searching for the seat of the fire, they found themselves claustrophobic between the hall's narrow walls. If a flashover occurred now, they would be trapped like blind mice.

Glenda forced the thought out of her mind.

Behind them she heard water brooking down a

staircase from the exterior aerial assault. It sounded like a waterfall. That lessened the chance of a flashover. Chapman probed the way ahead with short bursts of water, trying to locate fire.

"Hit it and see what sizzles!" he yelled.

Patches of flame appeared like a sun trying to shine through storm clouds. Dragging the heavy hose, their lifeline, the fighters knocked down flame, then cautiously advanced to the next enemy stronghold. Chapman soaked everything in their path; their boots splashed in water ankle deep. Any blaze left behind could build up and form a trap. The crew rotated on nozzle time. It took strength and balance to hold the nozzle. Occasionally nozzles had been known to escape and clear a room. A runaway nozzle fully opened was like trying to grab a rodeo bull. More than one firefighter bore suture scars where a nozzle caught him.

When it came Glenda's turn on point, she tucked the hose hard underneath her arm and wrapped her fingers around it so grimly that her arms ached. She racked the lever forward when flame jumped up in front of her. She gritted her teeth, planted herself against the hose recoil, leaned into it, held on, and to hell with this fire. To hell with the fire, where it belonged.

To hell with it.

A feeling of great ecstasy and excitement overcame her. While she might have doubted her abilities at times, as untried soldiers must, all her doubts suddenly evaporated. She had never felt such satisfaction in a job before. She could have laughed aloud from her jubilance, but laughing required more air, and all the air she had she carried in the tank on her back.

Something opened in front of her off the hallway. Closing the nozzle, Glenda felt with a gloved hand to determine the size of the breach. It seemed to be a doorway, but perhaps it was a section of collapsed wall. She entered what felt to be a small room. Smoke hung heavy here, thick and unmoving, unvented and impenetrable to the eye.

Somewhere in the room fire waited, smoldering in hiding. Glenda heard it feeding.

Suddenly it charged. The room flashed into brilliant white flame. Super heat drove her back a step. Crying out in surprise, she nonetheless refused to retreat. She learned in fire school that firefighters did not run.

Firefighters did not run.

She dropped quickly and smoothly to one knee like a machine gunner as plaster and smoldering rubble pelted her helmet, as tongues of flames licked their tips toward her. She found herself in the midst of an inferno.

"Glenda! Back it out!" Chapman shouted behind her, tugging on the line.

But Miami Beach's first female firefighter was meeting the enemy face-to-face, no quarter asked or accepted. Tilting the nozzle straight at the ceiling, she opened the pipe full out and held her ground. Flames licked and lapped and howled. The room proved to be small. The force of water shooting through the hose filled the room with a roaring cataract of water. It was like standing beneath Niagara Falls, which had, miraculously, caught fire.

While the fight seemed to go on for hours, in actuality it lasted but a few minutes. The flames seemed to moan as they died, replaced by thick clouds of smoke and steam. The room turned dark again in the smoke. Glenda took a deep breath through her SCBA, both of extreme relief that she had held out—and of extreme satisfaction. The firefighters had won another battle.

"Let's back out of here," Glenda gasped at last. "We're in a bathroom."

"Men's or women's?" Chapman inquired—and Glenda chuckled through her SCBA.

"That was some serious nozzle time, Guise," Chapman commented later, after second-in relieved first-in for mop-up.

It was like the first time in heavy combat, when the sarge came around afterward to say, "Hey, private. You did a helluva job." Glenda felt good. She felt like she could have walked on the last of the thin smoke emitting from the hotel. She had been waiting a long time for someone to acknowledge that she had done her job.

Never again, after the Saxony, was Glenda Guise to hear the male firefighters say she couldn't make it. Like most predominantly macho male groups, firefighters had a special if oblique way of letting a new member know he was accepted. Glenda knew she was accepted, had become a part of Station 2 and the fire department, when Irik or Barrett or any one of the others guys stopped pulling punches in deference to her being female. She returned as good as she received.

"Guise?" someone might quip. "Wanna make my Christmas? I told Santa I wanted a blonde."

"Not in this lifetime, big boy. The Hog's Head Chief is more your style. Go sit on *his* lap."

22

THE CALL CAME OUT AS A "DOMESTIC DISPUTE, WITH injuries." When the Rescue alarm sounded in the ready room, some of the firefighters were talking about little Manny, now a grown teenager, whom the crew of Engine 2 had spotted at a minor hotel fire. Manny ran up to his old friends from Station 2 and wrung their hands, like a brother just returned from a war or something.

"Where you been, kid?" the firefighters asked.

"Working, man, working. My old lady, the bitch, skipped out with some dude, and I ain't seen her cheap ass since."

Manny. With that haunting lost look in his eye.

"Stop around the firehouse sometime, kid. The guys ask about you sometimes."

"Yeah, man. I'll do that. Sometime."

Jim Barrett riding Rescue 2-2, first out after seven P.M., didn't hear the rest of the Manny discussion, if there was more to it. He jumped up and sprinted for the Rescue van. His crew arrived on the "domestic" ahead of police.

People filled lighted windows and lined the dark balconies of the high rise as the red Rescue van pulled to the curb. It must have been one hell of a family fight to bring out the entire building. A woman in curlers at an upstairs window tapped frantically on her pane and gestured wildly, pointing.

Barrett got out of the van and looked around. Everything appeared quiet enough now. The only person visible outside the condo was a thin man in his thirties slouching in the night shadows on the front steps. He wore pajama bottoms and slippers, but was naked from the waist up. He watched the Rescue team as he lifted a cigarette to his lips and drew smoke deep into his lungs. His hand trembled. He had eyes the color of new steel. Barrett noticed that when he drew near. They were as cold and dead as the eyes of a dead fish.

"It's upstairs," the man mumbled. "Second floor, second apartment on the left."

"Your apartment?" Barrett asked.

The man shrugged and looked away, smoking.

The Rescue men, laden with heavy medical aid bags, charged into the building. The police could handle the guy on the doorstep; EMTs took care of the casualties. Barrett and his two-man crew found the second floor deserted. All the doors were closed and apparently bolted from the inside. That was highly unusual. Curiosity seekers normally mobbed Rescue men at the scene of any tragedy. Not a door cracked, however, as the firefighters rushed down the dim hallway to the second door. Barrett's knock sounded louder than usual in the stillness.

He knocked again when no response came. "Fire Rescue!" he called out, making sure whoever might lurk inside the apartment knew the visitors were friendlies. "Fire Rescue. Does anyone need help?"

He tried the door. It was unlocked and swung slowly open, creaking like in some spook movie. The hair on the back of Barrett's neck twitched. He peeped inside, house, Chapman joked later. They checked out the living room and then the kitchenette and the single bedroom. Dirty clothing, old magazines, soiled dishes caked with rotting food littered the apartment. That was about normal for this neighborhood. The place was so cluttered, they couldn't tell if there had been a fight or not.

They came to the bathroom. As soon as Barrett pushed open the door, the strong, coppery odor of fresh blood assailed his nostrils. He switched on the light. The stool looked out of place with its frilly pink cover. Splatters of blood led across the matching rug to the shower stall. The stall door was closed. There was more blood splattered on it.

It looked like the end of the search.

Barrett slid the shower door open. He caught his breath and stepped back, bumping into Mogen and Chapman. They stared, aghast. Barrett thought *santerias* had been busy again sacrificing their chickens and goats. About ten gallons of bloody guts and flesh filled the bathtub. He identified a heart, the stomach, a large liver.

It must have been a goat, maybe two of them.

"What's with these people?" he wondered. "Miami Beach ain't ol' MacDonald's Farm and Slaughterhouse. The carcasses have to be around here someplace."

The only place left that they hadn't searched was the balcony. Streetlights and a pale, sad moon illuminated it.

"My God above!" Barrett breathed.

The carcass was not a goat's. Flesh on the young woman's legs from the knees down had been carved off the bones as from a Christmas turkey. She'd been gutted with a cut down her middle like a hunter looked down a short dark hallway that appeared to open into a small living room. The only illumination inside the apartment came from streetlights outside shining through the windows. It was a ghastly, greenish kind of light, like before or after a storm.

David Mogen and Neal Chapman were Barrett's partners. The three EMTs stepped cautiously into the hallway. All were aware that in the near-dark a fire department uniform with its badge might be easily mistaken for a policeman's. People in the heat of passion sometimes shot at cops. Captain Garcia's best friend, a Miami Beach cop, had been shot and killed in a situation similar to this one.

"Hello? Hello?" Mogen shouted.

There was still no answer, just the sound of the ocean breezes rattling a window shade or something against glass.

"Maybe we'd better wait for SWAT," Chapman suggested. "There was supposed to be a lot of screaming and fighting coming from this apartment. The guy had a gun."

During his years rotating back and forth from fire suppression to Fire Rescue, Barrett had entered his share of rickety hotels, rat hovel apartments, and crack houses. They always made him nervous; he never knew what to expect. While Ed Delfaverro as a SWAT medic had his gun, Rescue men had to be satisfied with bullet-proof vests, which they normally left in the van anyhow.

"I think this apartment belongs to the guy we saw down on the steps," Barrett said. "Wonder what he left up here for us."

"Do we really want to know?" Mogen whispered.

The hushed silence enveloped them. Feeling like house sneaks, the EMTs advanced into the apartment as a single unit. Like the Three Stooges in a haunted field-dressed a deer. There could be no rescue for this horror.

For a long few minutes the paramedics were too shocked to utter a word. They avoided each other's eyes, as though they were in the presence of some abomination that was not real if they refused to acknowledge it. From somewhere in the distance, wafting in on the fresh breeze, came the lilting sound of Christmas music: "On the sixth day of Christmas my true love gave to me . . ."

"Yeah," Barrett muttered. "Yeah."

Police arrived almost immediately after the discovery of the body.

"We got the guy downstairs who did it," a detective said. "He said he killed his girlfriend."

He looked at the mangled corpse. "I don't know if *kill* is the right word, if it's strong enough," he said. "Still, I suppose this one's not as bad as the guy who chopped off his girlfriend's head and was walking down the street carrying it by the hair and showing it to everybody."

The detective chuckled with the calloused sound of a hard man who had seen it all. "A real head case, that one," he quipped. "A real head case."

23

VANCE IRIK THOUGHT OF THE FONTANA HOTEL AND HOW the corpses had smelled like burnt steaks. He feared another fire like that unless he kept after the hotel arsonist and caught him. He didn't have to roll on fires, but he left instructions to be paged at home or wherever anytime a blaze erupted in any of the Beach hotels. Maybe he would get lucky.

In the middle of the night, yawning, with a cup of black coffee on the dash of his red sedan, he pulled up to a ratty strip hotel where a thin pall of smoke hung in the street. A few occupants in night clothing hung around on the sidewalk watching. It had been a garbage run, young Tim Daugherty on Engine 2 told him. A couple of Cubans had set their bed sofa on fire. One tried to burn up the other. The police took them downtown for questioning.

Daugherty smirked. "You oughta see these guys. It was a lovers' quarrel."

Another weird one like the naked guy. Obviously, it wasn't the hotel pyro, but it was an arson nonetheless. Irik was already awake; it would be an hour or so before he and his shovel could get into the burnt room. He drove to the police station to get a firsthand account of what happened.

"You're gonna like this one," Daugherty said.

Detectives had separated the Cubans and called for an interpreter. Neither Hispanic spoke English. The interpreter was a rookie policewoman who looked about eighteen but had to be at least twenty-one to join the force. A uniformed cop wearing a grin directed Irik to an interrogation room where the interpreter and a big-waisted detective hammered questions in Spanish at a woman wearing a low-cut black blouse.

Irik took one look and quickly backed out of the room. Obviously he was in the wrong place.

"Officer," he protested, "I'm looking for the Cubans from the hotel fire."

"Yes," said the policeman with the smug look.

"They're supposed to be *men*," Irik explained.

"Men," the policeman agreed.

"But . . . ?"

The cop guffawed. "You noticed," he chortled.

"That's a guy? He's got *tits*."

"What do you think—about a 38B cup?"

Irik took another look inside the room. The suspect had his legs crossed primly in a short skirt. He wore an across-the-heart bra that lifted his tits almost out of the blouse. He clutched a sequined purse in both hands.

"He wants to put on his makeup," explained the policewoman interpreter.

"Tell him he's pretty enough," the detective said, "but he could damn sure use a shave."

The detective greeted Irik and offered him a chair. Irik pretended not to notice the Cuban's tits, but it was difficult keeping his eyes from straying toward them. It was like talking to someone with a huge wart on his face and pretending not to notice it.

"This is the guy who set the fire," the detective

explained. "Best I have it figured out so far is that the other guy, the one without the tits, is this one's lover boy. Our Dolly Parton here got mad at him for using their money to score crack cocaine, so he set the sofa afire to show how angry he was."

"Must have been pretty angry," Irik muttered. He could have torched everyone in the hotel, that time of night.

The detective shook his head ruefully and observed, "This guy's got better cleavage than my wife."

The young policewoman blushed as the interrogation continued. In Spanish, she drew out details for a statement and possible prosecution. Irik noticed that the longer they sat there, the more the well-endowed Cuban squirmed in his chair. An expression of pain and embarrassment crept across his face.

"What's wrong with the guy?" Irik asked.

The detective looked at him.

"I mean, other than the obvious," Irik added. "Was he hurt in the fire?"

The policewoman rattled off a stream of Spanish, then translated the reply.

"He says he came to the United States in the Mariel boatlift. He was a criminal in Cuba because homosexuality is illegal. He's grateful that the U.S. is a lot more tolerant of homosexuals, but he says he decided he didn't want to be gay anymore."

Irik's gaze shifted. The guy was at least a 38B.

"He wants to be a woman," the interpreter concluded.

Judging from the appearance, he had a good start toward it.

"He's been taking high doses of hormones to get rid of body hair. I suppose it's obvious what else the hormones have done for him."

"He's got tits bigger than my Jewish mother-in-law," declared the impatient detective. "I'd marry him myself if he had a brewery and a recipe for pizza. So what else is his problem?"

"He's having a sex-change operation. He went to Mexico for the first half of his surgery."

The detective shuddered. "They cut off his . . . ?"

The interpreter blushed furiously. "They removed his penis," she affirmed.

Irik looked around. "Are we on 'Candid Camera'?" he asked.

"He was just released from Jackson Memorial," the policewoman continued, biting her lip to keep from laughing. "He had the second half of his operation done there."

The police investigator stared. He wasn't believing all this. "They gave him a . . . ?"

"They constructed a vagina for him."

"I'll have to see this," the detective decided.

"Is that what you want me to tell him?"

"Ask him what's wrong with him now. That was all we wanted to know."

She did.

"He says he has to go home now."

"We're not through with him."

"He has to go home. Where he had surgery he is starting to bleed. He needs a tampon."

"They gave him a period too?" the detective howled. "God, ain't medical science wonderful?"

Irik's was a crazy job as well as an interesting one. He possessed the ability to laugh at his own human frailties as well as at those of others. He finished the night's reports in his office behind Station 2. Afterward, before returning home, he stood at the window looking down Pine Tree Drive, with its colored Christmas lights sparking in the heavy pines. People. Who could ever figure them?

24

AT STATION 2, BACK BEFORE WIRY GENE SPEAR HUNG UP his helmet and turnouts for the last time, back when the sad little semi-orphan Manny hung around shooting hoops with the firefighters and asking questions about fire, Spear assumed position as the acknowledged resident expert in repairing old toys for new Christmases. Manny, with his hair falling all over his thin, drawn face, sat cross-legged at Spear's feet in the ready room while the fireman labored, replacing wheels on a wooden car. Nearby in a pile lay a doll without a leg, a Hot Wheels racer with its track broken, an electric train that had shorted out, some old record players, a few stuffed animals that required a little loving care. . . .

Though Spear was gone now, and the firefighters were occupied with their own Christmas—Barrett trying to select the perfect gift for his wife; Mogen joking about his circumcision: "No, it was three rabbis with a chain saw"; and Vance Irik serious about his hotel arsons—the firehouse never lacked for volunteers to fill Spear's void. Christmas toy tasks fell to Neal Chapman, Otto Ramirez, or to Stu Merker, who worked and made jokes and kept everyone laughing and interested in kids who ordinarily might not have had Christmas. Merker called repairing toys the firefighters' grand experiment in trickle down economics. There were never enough donations in new toys, but sufficient broken toys poured in during December to keep the firefighters busy repairing them on downtime and between alarms.

At other times, firefighters collected for the March of Dimes, conducted food drives for the needy, or hawked for charity donations at busy intersections. Firefighters sometimes took up collections on the spot for families whose houses and possessions had been burned. Back before Captain Jim Reilly died of cancer, he often rented motel rooms for fire victims or opened his own house to them.

In many ways, fire departments still occupied a special place, a place not shared by police or any other civil servants. Residents of a fire district looked upon the local station as *theirs;* the firefighters reciprocated in this feeling of community. Residents knew they could call upon firefighters for anything, *anything*—rescuing cats from trees, checking on a neighbor, finding a lost child, extinguishing a fire in a children's clubhouse—and the firefighters always rolled. Few other government agencies were as reliable. When there was no one else, when no one else cared, the fire department remained.

Christmas holidays proved especially lonely or trying for many people who had somehow gotten lost or cast aside. In the same way that the police had their regulars who called in a family fight every Saturday night or routinely wandered off from the nursing home and got lost, the fire department had its characters. They became a part of any station's folklore.

Hazel and Beth were two of these characters, who

always elicited a shake of the head or a sad smile at Station 2. Gene Spear knew them before he retired; Jim Reilly knew them. For years the city had threatened to condemn the run-down four-story apartment building in which the sisters lived. The only problem with that, Merker quipped, was that the city would have to condemn the inhabitants as well, since they were in worse shape than the building.

The rooming house stood on a corner at South Beach, like a scabbing sore on a diabetic's leg. Weathered yellow paint, old dandruff, scaled onto the weed-brown shoulders of the dying lawn. Whenever Spear entered the lobby on yet another pushout to check on the alcoholic spinsters, he felt he was entering a dark cave that should have bones scattered about. People cast onto the ratty sofa and chairs resembled mummies in the dim light. Their eyes creaked at the firefighters. Merker suggested they routinely check pulses. It would save the government a lot of money mailing out Social Security checks to people who had been sitting up dead since last Easter or Groundhog Day.

Hazel and Beth regarded the fire department as their personal service. They called whenever one of them had a headache or the flu or a hangover. Until that one fateful morning in December, they seldom had anything *really* wrong with them that a few AA meetings wouldn't clear up. Each day for them was like the first day of a new bottle. While they were the firehouse's most faithful patrons, they never recognized even the firefighters like Spear or Barrett, who had responded to their calls numerous times.

"It's the sisters again," the dispatch always announced.

The sisters. Everyone knew who they were.

"Those old biddies are going to die up there one day," Jim Barrett predicted.

On that day when something proved really wrong on the sisters' fourth floor, Gene Spear pounded on their door. Since the spinsters were hard of hearing, their TV ran full-blast, knocking cockroaches out of hiding. Only their neighbors being equally hard of hearing kept down police complaints.

Hazel finally came to the door. The face of an anemic gerbil peered through the crack permitted by the security chain. "What do you want?" she demanded, blinking.

"Hazel, look at us. We're the fire department."

"What?"

"Go turn down the television. Hazel, turn down the TV."

"No. No. Beth don't have TB. She just ain't feeling well. Who are you?"

"The fire department. Did you call in, saying someone's sick?"

"Thick, you say?"

"Fire department!" Spear shouted, loud enough to shake plaster dust from the ceiling.

"Oh. Fire department. Why didn't you say so? I didn't recognize you. A body can't be too careful these days."

"Yes, ma'am."

The firefighters entered another musty cave. It smelled like last year's dirty underwear left soaking in a vat of gin.

"Beth and me, we always count on you boys," Hazel said. "Your mothers must be proud."

"Where is Beth?"

"No, no, there's no death. Beth ain't been feeling well, is all. She didn't wake up this morning. I let her sleep and decided to call you boys."

She led the way to the bedroom, but remained outside the door. "Poor Beth has been awfully quiet," she whispered.

Beth could have been Hazel's twin gerbil. Her little rodent's face glowed pale in what light penetrated the drawn shades. She lay in bed with the covers pulled up to her chin and her hands arranged peacefully outside the covers. She felt cold and stiff to the touch. Spear noticed tears easing wet streaks down Hazel's poor, mottled old face even before he spoke.

"She's gone, ma'am."

"No she aint. She's right there."

Spear escorted the tottering woman to a chair in the other room.

"I—I don't know what to do with her," Hazel said.

Spear knelt next to the old woman's chair. He was good at fixing toys and other broken things. "We'll take care of it for you," he promised.

Gene Spear left the fire department before the next Christmas. After that, Glenda Guise or Otto Ramirez or Neal Chapman riding Rescue answered the calls to South Beach and the fourth floor where Hazel survived alone.

"Hazel, it's Rescue."

"Blue? God, I can't stand blue."

One morning Chapman pounded on the door and no one answered. The ancient hotel manager lugged the key upstairs and let the paramedics in.

"If you're burglars," Hazel's thin voice piped from the bedroom, "there ain't nothing worth stealing. You better leave. I've called the fire department."

"Hazel, it *is* the fire department."

"Apartment? Yes, my apartment."

From her looks, Hazel might have been the world's oldest living human being. She blew her nose on her pillowcase and blinked at the intruders.

"You're sick," Chapman said. "You want we should take you to the hospital?"

"Pistol? You don't want to shoot an old woman, young man."

"No, ma'am."

"Heavens now. You're the fire department. I just have a bad cold. I needed someone to get me a glass of water, and I just didn't feel like getting out of bed. There's just no one else, you know, now that Beth is gone."

"Ma'am, we'll get you a glass of water."

Hazel and all the other old ladies at Christmas.

"We'll get you a glass of water," Chapman said.

25

THERE WERE CERTAIN THINGS FIREFIGHTERS REMEMBERED about the big fires, and certain things that simply melded with other images and recollections until what remained were memories of fire in general. It was that way with Rod Harris. He remembered fire with no particular fixed location or time or feature, just *fire,* and then he remembered Manny's Restaurant.

He remembered that videos of the Manny's Restaurant fire were used in fire school to teach procedure to probies; he had thought he was going to be burned alive.

Every piece of firefighting machinery the department owned rolled on Manny's. The building was two stories, with Manny's Restaurant below and a dry cleaner on the second floor. It was constructed of Florida pine because pine repelled insects. Pine also seemed to attract fire. Resin in it made it light up like kerosene. Harris smelled the burning almost as soon as his engine clanged out of Station 2 onto Pine Tree. The roof was well-involved, gushing smoke, by the time Meridian Avenue filled with apparatus, helmeted firefighters, cops, and crisscrossing hoses.

Heat was so intense it seared exposed flesh a hundred feet away. People ran screaming out the front door into the waiting arms of firefighters. Two paramedics worked over a pile of smoldering rubble where a wall had crumbled onto the sidewalk. They slung bricks and smoking lumber, trying to reach an elderly woman upon whom the wall had fallen. Finding her, they pulled her out by her arms while she screamed and flailed about.

She later died in the hospital.

Pressure buildup was already popping windows, hurling shards of glass like shrapnel. Fifty feet away, on the opposite side of the street, a police officer dropped as though he had been shot. He smeared the sidewalk with bright blood as he thrashed about in pain. A fragment of glass had ripped open his chest.

The cop lived.

Hazardous as it was, the fire officer sent Rescue teams into the inferno to search for trapped survivors. As a member of one team, Harris penetrated the heat, the choking smoke, the rubble, and the chaos of a fire that was really cooking. Somehow, after fighting his way to the large kitchen, he found himself separated from the rest of his team.

Almost the same instant he realized this, he heard a

deep rumble, as if from the cavernous throat of an active volcano. The entire building trembled, almost throwing the firefighter off his feet. He thought the roof and the second floor were coming down on top of him at any moment. His first instincts told him to flee, to save himself and get the hell out.

Second thoughts, however, his firefighter's sense, sent him stumbling blindly forward through the smoke. A firefighter never quit—not when there might still be victims to rescue. Quickly, he searched the kitchen, the walk-in freezers, the adjoining rooms. He couldn't know that everyone had escaped the building until after he searched his assigned area. Simply because no one remained to be rescued made him no less a hero for trying.

Another cavernous rumble shook him on his feet. Assured that he had overlooked no survivors, Harris felt his way back across the kitchen. Suddenly, an explosion rocked him. A brilliant flash-bang from overhead hurled flaming rubble. For that moment the world around him—at least *his* world—collapsed in flames.

Desperately, feeling deep in his guts that he was a goner, he threw himself underneath a stainless steel table just as the ceiling collapsed, showering flames and rubble. Even through his SCBA the heat was so intense he could barely breathe. It gripped his lungs while terror took control of his heart. He knew, he *knew*, that he was about to be buried alive, then incinerated.

The steel table was one of those long utility ones that stretched nearly from one end of the kitchen to the other. It saved his life. He rolled and crawled to the opposite end of the table. Debris piled up on the table and around it. The firefighter felt as though he were cooking. Maybe just medium rare, but cooking nonetheless. A man could sustain that kind of heat for a few minutes at most.

It was there, momentarily trapped in the inferno, that Harris felt a sudden change in his outlook on life. If he lived, he swore he would never again take life for granted. With that thought sustaining him, he found an opening in the burning debris at the end of the table and burst through it.

A minute later he fought clear of the flaming building and sucked in the sweet taste of fresh air. Afterward, he always remembered how high he felt at merely being alive.

Firefighters remembered things like that, which were significant in their lives. Harris always remembered Manny's Restaurant.

26

STU MERKER DECIDED THAT IF HE EVER WROTE HIS autobiography, he would title it "Rescuing Mr. Jones." Mr. Jones characterized the Christmas Silly Season. For over ten years, each December, he came down from the North, New Jersey or New York or somewhere, and rented the penthouse suite at the Fontainebleau. He weighed about five hundred pounds, "maybe twelve pounds heavier than a Cadillac," as Merker put it. And for each of those ten years,

Mr. Jones knew he was dying of congestive heart failure.

Whenever he went out for an evening and consumed enough food to cater a wedding reception, washed down by a keg or two of Diet Pepsi, and thereafter couldn't catch his breath, he dialed 911 in a panic.

"This is Mr. Jones. I'm dying."

"It's finally Christmas!" someone shouted.

"Rescuing Mr. Jones..." intoned the knowing dispatch. "Rescuing Mr. Jones... He's dying—again."

"He's not dying," declared Tim Daugherty with an expressive sign, "but he's damn sure killing us."

The logistics required to maneuver a quarter ton of human beef from the penthouse to the lobby had been worked out long ago, on Mr. Jones's first December visit; it became continuing testimony to the ingenuity of the fire department and the cooperation of the hotel employees. Everyone knew the drill, the enduring details of which passed from generation to generation like an heirloom everyone found necessary to his quality of life. Mr. Jones tipped well and spent a lot of money, which lent incentive for hotel management to ensure his continued long life.

"Why couldn't he take a first-floor room?" Otto Ramirez complained. "It'd make a lot more sense."

"Any inconvenience caused by Mr. Jones's... ah, condition," sniffed the hotel concierge, "is more than compensated for by his generosity and good manners."

The drill started with the elevator waiting and a pathway cleared from the elevator to the front doors. Rescue never bothered bringing in a stretcher. Mr. Jones was too big for a stretcher when he was ten years old. He was too big for the door to his penthouse, having to enter and exit sideways. He filled the elevator. One Rescue man rode down with him. The others waited for the next elevator.

"Good day to you, gentlemen," the concierge greeted upon Rescue's arrival. "I take it your Silly Season is beginning."

"Day One," Ramirez replied. "It's show time."

Mr. Jones took up the end of a sofa facing bright, open windows overlooking the ocean far below.

"I'm terribly sorry about this, gentlemen," he apologized, wheezing desperately. A voice as small as his belonged on the circus dwarf, not the circus fat man. He was so pleasant a man that it was difficult not to like him. "I don't understand what it is about your Miami Beach weather. Oh, it's a lovely sun and all. But I feel absolutely horrible. I began feeling dizzy. Light-headed. The entire room started spinning and I sat down."

His illness wasn't all in his head. He did have a heart problem.

"The chair?" he asked. He knew the drill too.

Ramirez smiled. "The chair," he agreed.

From somewhere, the hotel management had acquired the strongest chair in South Florida. It was an ugly straight-backed kitchen chair constructed of oak bolted and notched. A Budweiser Clydesdale could have sat on it. Whether the chair would hold or not was never brought into doubt; the doubt always came in lifting the behemoth off the sofa or bed or floor and balancing his broad butt upon the bottom of the chair. Sometimes it took four firemen and two bellhops and help from the kitchen to lift him while the concierge stood stiffly by, making sure Mr. Jones never felt violated.

Mounted finally on the chair, his body folded and flabbed and draped around it, he resembled a fat man with four short stubs protruding from his bottom half. After everyone caught his breath and regained his strength, the firefighters picked up Mr. Jones, chair and all, and laboriously maneuvered him out the penthouse door, down the elevator, and into a waiting ambulance.

"I'm so sorry. I'm so sorry," Mr. Jones apologized all the way down.

"You'll be as good as new in no time," Tim Daugherty assured him.

"Sir, I was *never* as good as new."

Mr. Jones was only about forty. He was digging his grave with a spoon. But maybe that was the way most people went out, Otto Ramirez conceded—dying slowly but relentlessly from a lifetime of excesses.

Stu Merker stood outside in the December sun and peered upward at the penthouse windows.

"Wonder how we're going to get him down from up there when he finally dies and we can't sit him on a chair?" he mused.

27

FIRE AND ARSON INVESTIGATIONS, ESPECIALLY DURING THE Silly Season when fire stats climbed like mercury in August, kept Vance Irik running. So far he had dug up no clues as to the identity of the pyromaniac setting hotels afire. On top of that, the city expected him to keep up on his other duties.

In Miami Beach, as in most cities, the arson investigator was also the fire prevention inspector and the fire protection officer. This meant that the government not only granted him uniquely broad powers to investigate suspicious fires, giving him the right to enter any fire site for inspection and investigation and to subpoena all records and persons associated with the fire, but also held him responsible for a number of other duties as well.

As fire prevention inspector, he conducted periodic inspections of public buildings and enforced the fire code—rummaged around in steamy restaurant kitchens checking for such as oily rags piled next to ovens; looked over condos and apartment complexes for violations of fire lanes blocked or lack of fire alarms or functional extinguishers; tapped out-of-date extinguishers in theaters, clubs, and restaurants and made sure public exits were kept unlocked and unblocked and that no business operated over maximum occupancy.

As fire protection officer, he issued licenses and permits for various occupancies and functions, was responsible for fire prevention education and fire service training, kept fire loss statistics, and regulated fireworks, flammable liquids, liquefied petroleum gas, and other hazardous materials.

All of which was necessary, but time-consuming. And mostly boring as hell. It took him away from his real passion—arson. Usually, a local fire department or state fire marshal was responsible for the initial determination of arson, fire causes, and the collection of evidence, while the criminal investigation itself fell to the police. Although officially that was the way it was on Miami Beach, the police included Irik—and

Irik included himself—in any arson probe, from the fire scene to the arrest and interrogation. After all, Irik determined when and if a crime had been committed in the first place.

Arson investigation, in dealing with crime and the criminal mind, often led him to sordid secrets. He found that everyone—upper class, lower class, rich or poor—had their dark little closets in which they hid family secrets. Policemen and firemen, in doing their jobs, often barged through locked doors; they saw it all. It had to be embarrassing when some tragedy exposed people for the authorities to poke through and make judgments on. Irik hated and at the same time was fascinated in looking at other people's hidden skeletons. Skeletons told him so much he didn't want to know about human nature.

He never forgot the looks on the faces of people made to confront the exposed skeletons of those they loved. He remembered David Mogen working Rescue 2 making a more or less routine run to an apartment building on Washington Avenue. Some tenant in the apartments hadn't been seen in several days. Often, if the tenant were elderly, a call like that meant a dead body in the room. This tenant, however, was a young kid, maybe twenty, who had just left home to start out on his own.

The apartment manager unlocked the kid's door to let Rescue men enter. Irik, who had been nearby on a follow-up investigation, stopped by to offer what help he could. Whatever their specific assignments, firefighters remained generalists who assisted other firefighters. Mogen led the way into the apartment. He groomed his mustache carelessly as he looked around the young bachelor's pad. It was cheaply furnished and a bit messy, but not trashed. It resembled any teenage boy's room at home, with Michael Jordan posters on the walls, other posters of rock groups, and centerfolds from *Playboy*.

Mogen opened the closet door. He stepped back. "Call the police," he said.

Another few weeks perhaps and the kid would have been his own skeleton in the closet. The boy hung by his necktie from the closet clothing rod. He was naked except for the necktie, and had pitched forward a little on his bare toes. Neck garroted by the tie, his face was an awful overripe fruit, discolored and mottled a shade of bruised-blood purple. Tongue like a dead bloated gila lizard protruded through hideously swollen lips.

The kid's family wanted to believe his death was murder or even suicide, over what it really was—an accident. Irik had seen similar deaths before. Young men, mostly teenagers, had discovered they could intensify orgasm in masturbation by cutting off the oxygen supply to the brain at the crucial moment. Sometimes, like now, they cut off too much oxygen, passed out in between strokes, and literally hung themselves. The kid's mother and father waiting outside stared like deer caught in a car's headlights.

"I hope the hell it was worth it," murmured a hard-nosed detective who arrived to take a look and make a call on the cause of death. "Died jacking off. How'd you like to have that in your closet?"

Irik had seen a lot of skeletons over the years that he wouldn't want in his closet.

Now he had another one of those "skeleton" calls waiting for him when he checked back in with dispatch after hanging a surprise early morning code inspection on a dinner club on Collins Avenue. The inspection had resulted in a dozen violations, ranging from out-of-date extinguishers to locked fire doors. The owner was a tubby man who sweated healthily, squirmed and cleared his throat and scratched himself.

"Government is destroying the businessman," he said. "Back in the old days in Philadelphia we could make accommodations."

What he meant was bribes. There were times in the past when being a fire inspector was an assignment so choice that firefighters fought over it. Every firetrap and slum owner in a city slipped the inspector a hundred or two periodically to overlook violations. Not anymore. Irik ignored Tubby's hinting.

"Pass code," he advised, "or the city can close you down as tight as a dog's butt."

Dispatch had a message for Irik: "Lieutenant, Engine Three is at Fisher Island at the scene of a minor house fire. They want you to take a look at something hinky."

Irik caught the ferry over to the Miami Beach sub-island off South Beach where coral and white condos punctuated the city's most exclusive residential district. Fabulous mansions shined like jewels along the bright necklaces of canals and waterways. Farther north, Fort Lauderdale liked to call itself the Venice of America, because of its waterways, but the Beach ran a close race for the title. Yachts and Mercedes were as common on Fisher Island as Nissans and Chevys in North Miami.

The address where Irik dismounted was a two-story modern with a BMW in the drive, a lawn manicured and laundered each morning, and a pair of noblewomen—apparently mother and daughter—clutching each other near the Fire Rescue van. Irik's eyes took in the pumper and two police patrol cars at the curb before his gaze settled momentarily on a fire-scorched window on the house's second floor. It peered at him like a black eye. One of the firefighters told him the blaze had been confined to a single bedroom.

"The flames were red and orange with gray smoke," the fireman briefed Irik. "The smoke turned black when we turned the booster hose on it."

Nothing unusual about that. Nothing "hinky."

"That's the homeowner and her daughter. There was another young woman, a friend of the daughter's, who apparently inhaled a roomful of smoke. Rescue Three transported her to Jackson Memorial. The fire started about seven A.M. Seems the two girls were trapped in the bedroom together for a while. You might want to take a look at that bedroom, Vance."

As the tall fire investigator approached, the mother's steely aristocratic eyes snapped at him as if he were a cockroach. Irik squirmed and tugged at his collar; he couldn't help himself. Mother seemed more upset than possible over a minor fire. Mother had an attitude that reflected itself in a cold, defensive response to his routine questions.

"My daughter is most concerned about her friend," Mother announced, shielding her daughter with her own body. "Why don't you conclude whatever it is you conclude and permit us to go about our affairs. The insurance will take care of damages. I don't understand why you were summoned."

Irik reacted. He had piercing eyes of his own.

"How did the fire start?" he asked, getting to the point, his voice coming out unexpectedly gruff.

Mother's stare shifted quickly. She looked at the house, at her feet, at the fine blue-veined hands that tumbled over themselves in front of her. She looked everywhere but at the investigator. Her nervousness attracted a slight frown from Irik.

The daughter appeared equally agitated. Her eyes kept banking like balls at a pocket pool tournament. A younger replica of her mother, she was about twenty, slender, fine-featured, of that economic class normally accustomed to treating public servants like servants.

"We were sleeping when the fire started," the daughter mumbled, her eyes still averted.

Okay.

"The girls are very close," Mother added. She met Irik's eyes then, as though challenging him to make something of it. "They are like sisters. They even sleep in the same bed."

Okay.

The women hesitated. They wouldn't look at each other either.

"Well..." the daughter began, mumbling. She began several times. "Well, I accidentally knocked over a candle in the bedroom. It didn't look like a big deal at first, you know. We tried stomping it out ourselves, but the fire... it—it spread so fast...."

Okay. A candle. It was the Silly Season.

"Take a look inside the room," a policeman advised Irik, aside.

Armed with a Polaroid camera and his ever-present shovel, directed by the thick odor of charring and smoke, Irik climbed upstairs, conscious of mother and daughter staring hard after him. It puzzled him, their demeanor, until he stepped through the bedroom door sooted by smoke and scorched at its base and looked around.

First of all, treating the room as a potential crime scene, the fire detective carefully noted the overall general appearance.

The bedroom was large and expensively furnished, although two dressers, a nightstand, and a bookshelf now resembled stumps of trees left in the wake of a forest fire. Walls bore the tongued black signature of the fire. Shreds of fire-blackened curtains hung over windowpanes shattered by firefighters venting the flames. Expensive shag carpeting around the bed lay melted and carbonized to the floor, but the queen-sized bed itself had survived in remarkable condition. Only the base of the bed—its sheets, spread, and frame—had been incinerated. Water and soot had collected on the bed's surface, but the blaze itself had been kept at bay.

Frowning, puzzled, Irik squatted next to the bed. His fingers traced a series of wax outlines melted into the carpet. It made an interesting ring around the bed. It was like the bed had been a shrine to which candles were lit. It hadn't been just *a candle;* it had been dozens of them.

Irik began to comprehend as his eyes next found lengths of soft cotton rope knotted to both the bed's headboard and footboard. From each length of rope dangled studded leather wrist manacles of the type advertised in underground S&M (sadomasochistic) publications. From the bed he then recovered a penis vibrator the size of a stallion's cock, and a vial containing residue of white powder. The scorched dresser drawer contained another thin vial of fine crystalline powder. Irik had seen cocaine before.

There was nothing, however, to indicate the girls, like comedian-actor Richard Pryor, might have been freebasing the stuff and caught the room afire. Obviously, the girls were involved in something more *personal*. The sleuth looked up and away from the room, out the window at the bright Florida sunshine. The cities of South Florida were always so bright and white they hurt the eyes. Few tourists suspected the dark secrets that lurked beneath.

The young girl below in the yard, sharp-featured as she was, appeared on the surface as bright and white and innocent as a Florida city. Mother, to her horror and shame, had apparently caught her first look at the darkness beneath her daughter's sunshine. Irik felt he himself might have gone on the defensive if he had just discovered his daughter involved in a lesbian S&M relationship. Responding to the fire and the girls' frantic cries, Mother must have charged to the rescue and found one of the girls, naked, strapped to the bed while flames started by an overturned candle blazed around them. The fire had burned through the family closet—and there the skeletons were.

It wasn't in Irik's basically gentle nature to point the skeletons out once they were uncovered. Downstairs, the expression on Mother's pale face—a mixture of guilt and dismay—further confirmed Irik's analysis. He needed only a brief look at the daughter's girlfriend and perhaps a statement from her to conclude his investigation. Let the police handle the drugs; they were out of an arson investigator's jurisdiction.

"Sir?" Mother asked, her lips drawn tight and thin. She hesitated.

"Sir, I—I . . ."

"You don't have to say anything," Irik said, to which the woman responded with a sudden and almost overwhelming look of gratitude.

The daughter had already taken the BMW to the hospital to check on her friend. Irik discovered her at her friend's bedside. The friend was sedated, sleeping. She was very pretty, with long brown hair and eyelashes a mascara model might have killed for. Irik judged her to be about eighteen or nineteen.

He examined the bruises on her wrists from the manacles. Getting her out of bondage and out of the room to safety must have been nip and tuck against the rapidly spreading flames. While the daughter avoided his eyes, Irik decided on his ruling as to the cause of the fire: accidental. It required no further investigation. Let the family sort out its own skeletons in its own closet. After all, it was Christmas.

"The doctor said she's going to be fine," the daughter mumbled. "It was just so much smoke."

"Good."

"Look. I want to explain—"

Irik's palms flew out. "No need."

He didn't want to hear it. He turned at the door. "You have to be very careful with candles," he said.

28

THE SILLY SEASON KEPT FIRE RESCUE CONSTANTLY ON THE move. Most of the pushouts were back-to-back garbage runs—minor auto accidents, heart attacks, falls in the bathtub, home poisonings or burns, an occasional suicide, some tourist on the beach stung by a jellyfish. Captain Luis Garcia, commander of Station 2 Rescue, kept two three-man teams on twenty-four-hour readiness. The teams rotated on first out, with one team taking first calls for the first half of the shift while the second team ran backup, then switching teams for the second half of the shift. The holiday season was so busy that the teams sometimes lost track of who was first out and who second.

Soft-spoken John Creel, working Rescue 2, first out, complained good-naturedly that his crew ran a pair of garbage calls back-to-back and missed a quality construction accident to Rescue 2-2. He felt first out should have priority on the good calls, with backup taking the garbage runs. When fire departments were first organized, rivalry between fire companies proved so intense that a company's swiftest runner took off from the station as fast as he could when the alarm sounded. His job was to reach the fire plug at the fire site and hold it for the arrival of his own company. Vicious fistfights sometimes broke out over possession of the fire plug and over which company had legitimate rights to fight the fire.

The nature of firefighters being what it was, cherubic Steve Hoffman riding 2-2 couldn't permit Creel's complaints to go unanswered. He acquired the crying mechanism from one of the donated dolls that could not be repaired. Each time Creel's Rescue 2 took first-out alarm, Hoffman blasted him over the radio waves with his baby's wail. "Cry like a baby, Creel," he laughed. The first time the baby cried over Captain Garcia's desk radio monitor, Garcia fell back in surprise. His first inclination was to blast the offender.

But then Garcia, short and stocky, with his hair turning fatherly white, shook his head and smiled slightly. It was just a friendly feud, all in fun. Let them go at it. It wasn't terribly professional, but it helped reduce the tension of back-to-back alarms. Besides, he knew the paramedics turned serious when it was required. Other states and cities used Miami Beach Fire Rescue as a model from which to build their own paramedic corps.

Christmas made Garcia melancholy. A Cuban grown middle-aged on the Miami Beach Fire Department, he had watched from this same station house year after year as city workers came each December to stretch the same bright lights in the pines that lined

Pine Tree Drive. He felt he owed much of the good things in his life to the fire department.

For over two decades he had hurled his body into high-rise fires, and helped create Fire Rescue along the way. Firefighting was a young man's job; Garcia no longer felt so young. His own son had now applied for the MBFD. Sometimes, especially around Christmas, Garcia remembered his career as though it were a VCR movie he could replay at will. After twenty years in the same firehouse, the station became as much home as *home*. The firehouse held such memories for Captain Garcia. He stared into the replica Christmas tree on his desk.

The son of Cuban refugees, Garcia was among the first wave of Latinos to make their mark on South Florida. City government and its services held out longest against the invading "foreigners." In 1966, for example, there were only two cops on the Miami Police Department who spoke Spanish; one of them was Puerto Rican, the other was a quarter-breed Cherokee from Oklahoma. Twenty years later, most of the department and the surrounding smaller departments required a knowledge of Spanish in order to be hired.

Fire departments held out even longer than the police. The turnover among firefighters is astonishingly low, attrition coming about mainly through retirement. When Garcia applied for employment with the Miami Beach Fire Department, he had his doubts about being hired. Fire departments were good ol' boy networks, enclaves of white males who guarded their territory jealously. The name Garcia broke through that shell.

Captain Jim Reilly, then a firefighter with Engine 2, was assigned as Garcia's mentor. The red-haired Irishman studied the Latino from the top of his blue cap to his black shined shoes. Then he grinned that infectious smile that never failed to warm the air around him.

"Cuban, huh?" the Irishman noted. "Well, come on in here, Garcia. Ain't nobody going to eat you. Way I always figured it, ain't no Cubans or Shanty Irish or Polacks in turnouts and boots. You ain't no Cuban no more, Garcia. You're a *firefighter.*"

Fifteen minutes after Garcia reported for duty, Reilly called for a fifteen-minute coffee break. Reilly was like that. He taught at fire school. When the burn building heated up and the probies got to sweating in their heavy turnouts, he often called for a break, and they all rushed to a nearby lake for a quick swim. The only thing the redhead took seriously was firefighting, Rescue, and his family.

"He's one hell of a firefighter," other flame eaters said of him. "He'll rush in where even angels fear to get their wings scorched."

Garcia hoped the same would be said of him when he retired, that he would have left his mark on the younger station family as the older family had left its mark on him. He would miss Firehouse 2, like Gene Spear missed it. It would be like leaving home all over again. Year after year it had become a part of his life. It defined who he was. The alarms, the camaraderie, the courage, selflessness, and dedication of the robust and raucous flame warriors of Station 2. That was all a part of who Captain Luis Garcia really was.

Jim Reilly died of cancer while he was still on the job. He had been a relatively young man, as young as Garcia was now. Reilly died at home and would therefore never have to leave home.

Garcia stood from his desk and walked to the window, from which he peered down lighted Pinetree as down his memory. Winter, with its occasionally brown light, choked the sun, just a little, just for a few days. There had been a cold front overnight, bringing a record low of fifty degrees. Lowering clouds, rare for this time of year, blotted out the sun, threatening rain.

It always seemed to rain or was about to rain whenever Garcia's long memory resurrected the ghostly images that prowled his brain and illuminated his own mortality. The young—Glenda Guise, Neal Chapman, Otto Ramirez, and the others—the indestructible immortal young, were rarely aware of their mortality. Death and decay, disaster and accident— the young firefighters witnessed it around them, but it remained somehow remote. It had not become personal, as it would in later years. It had not touched them with the truth that man had far more in common than their differences. The personal aging process brought with it the awareness that the most common thing about man was his death and decay.

Captain Garcia had had his successes, his triumphs, perhaps even more of them than the average man. What was it about Christmas and gray days that sank him into that dark abyss of memory in which he and every other firefighter, every paramedic, tossed the human debris of the absurd, the grotesque? Such images surfaced, as they surfaced now, at the most unexpected times, the most inappropriate, as a homeless leper at a wedding reception.

The Cuban-American stared into the Christmas lights that lined Pinetree, but his eyes recorded nothing. He remembered . . .

"It's . . . it's my husband Harry. He went to the bathroom over an hour ago. He locked the door—and now I've been pounding on it and he won't answer."

Fire Rescue pounded and then kicked in the door. There Harry was, just as his wife declared. Harry with his pants down sitting on the stool. What kind of indignity was it when a man died of a heart attack while straining to take a shit? The gods had a sick sense of humor.

Elvis died the same way, on the stool. Dead on the stool with his pants around his ankles. Elvis and Harry.

"My husband? Oh. He and Elvis died taking a shit."

"The crazy little fruit. Bastard owes me rent money. He's been in there three days and he won't come out. Smells like he hasn't taken the garbage out either."

Rescue kicked in the door. It was the only way. The crazy little fruit, the bastard. He cheated his landlord out of a month's rent by blowing out his own brains with a .357 Colt Python loaded with 110-grain solid lead bullets. The neighbors said he was a fruit all right, but he tried to build a decent little life for himself in a shitty, rat-infested apartment. Things just went wrong somehow. He sat on a stool at the kitchen bar and wrote a long note to his parents back in Missouri, apologizing for everything.

"I know I've been a big disappointment to everyone . . ."

Then he blew himself off the stool with the Python. Blood and bits of bone and brain splattered the walls like an obscene work of modern art. Ants nibbled away at the edges of the stains, while worms gnawed relentlessly on the corpse.

"Okay, okay, who's responsible?" demanded the landlord. "Who's going to clean up this mess?"

Adult movie theaters with marquees proclaiming *Sex! Sex! Sex!* garnished Collins Boulevard. Next to them were adult bookstores, while around the corners huddled cheap bars and pawnshops. They were the underside of the glitter on Miami Beach. Sometimes, hookers waited on the street to snatch up men as they exited, primed from two hours of *Debbie Does Dallas* or *Deep Throat IV*.

On the screen, three or four couples writhed sweating and groaning and panting, coupling with everything except the coffeepot. Isolated shadows in the audience section sat alone with coats or paper bags or something over their laps.

"He's been in here for hours," the usher explained to Fire Rescue. "He's just sitting there with his eyes open. Not blinking or nothing. He came in at first show, about ten this morning, and now it's five and he ain't moved all day. He ain't moved. I tried to talk to him, tell him to move on, but he ain't moved. . . ."

Garcia shook the man's shoulder. "Hey, fella . . . ?"

This time the guy moved. The astonished Rescue man sprang back as the man's knees flexed underneath him and his shoulders slowly contracted. His head lolled to one side.

"Jesus God Almighty! He's dead!" the usher shouted as he bolted up the aisle. Three or four patrons jumped up to look. The lights flashed on.

Patrons, Rescue men, and the usher from up front peeping around the curtain watched mesmerized as the corpse in the movie theater slowly toppled from its seat and wedged on the floor between the rows. Its muscles contracted like old rubber bands. They jerked the corpse into a weird kind of slow-motion dance. Finally it stopped dancing and drew itself into a twisted, hideous caricature of a man.

The guy died while watching *Sex! Sex! Sex!* and silently masturbating in the darkness.

* * *

"Rescue Two, Code 45—Twenty-third Street and . . ." The dispatch's voice faltered. *"Stand by,"* she said.

Code 45 designated a DOA, a dead body.

"Rescue Two, the address we have is Twenty-third Street and Pine Tree. In front of the firehouse."

"She has to be shittin' us," Merker decided.

The station emptied. Astonished firefighters bunched around a dead man sprawled in the gutter across the street. Blank eyes stared into the morning sun. Young, some crackhead or mainliner emaciated from a steady diet of drugs, he appeared to have overdosed while staggering down the street. His heart suddenly exploded in his chest. And there he lay dead beneath the pines.

"It's not enough that Rescue goes to them," Garcia commented tersely. "Now we have them coming to us."

Captain Garcia stood at the window overlooking the Christmas lights lining Pine Tree. No rain fell outside, but down Garcia's memory it was raining. Raining hard, drumming against the windowpanes of his memory. Why was it he recalled such things? He must be getting old.

His soul flinched as Hoffman gave Creel another bleat over the air with his doll's baby cry, a painful reminder of a real baby that would never cry again. After all this time, Garcia closed his eyes and saw that mother's eyes again. They were filled with trust and with confidence that the skilled paramedics at the fire station could fix things, could fix her baby.

It rained the day she appeared. She appeared unexpectedly out of the rain with the sodden wind whipping at her light sundress and at the tattered old-fashioned shawl drawn around her head and shoulders. Cradling something small in her arms, wrapped in the tail of the shawl, she entered the fire station, walking heavily. She plodded in and, full of purpose, without looking right or left, made her wordless way to the ready room. The Boar's Head Chief scowled down at the intruder with its fierce glass eyes.

Otto Ramirez stood up, stunned by the look of the woman and the strange expression that gripped her face. Maybe Irik was there too, and Barrett. All in the ready room stood up and stared at the creature whose face revealed the childish unfinished features of a Mongoloid. She was obviously retarded. Slowly, she unwrapped her tiny bundle to expose a baby that had apparently stopped forming in the womb before it was ejected. It too was a Mongoloid.

"Fix," the poor woman entreated, thrusting the baby toward the firefighters. "Fix."

Like it was a broken Christmas toy that needed a new battery or a plastic arm mended. Captain Garcia looked at the baby; he looked at the mother with raindrops and teardrops mixed on her cheeks. Something vital slowly drained out of her eyes when he explained that the baby was dead and that there were some things firefighters could not fix. Something happened to her inside. She assumed the resigned air of an animal puzzling over its stillborn. Again and again she lifted her baby's tiny hand and let it fall, let it fall, trying to bring it back to life because death was something incomprehensible.

Garcia waited with the retarded mother in the ready room until the medical examiner came. He couldn't bring himself to let her wait alone. She huddled on one end of the sofa clutching her dead baby. Garcia occupied the other end. They sat in silence for an hour. The rain came in a thundering tropical deluge.

"I am so sorry," Garcia murmured once, "that I could not fix your baby."

It rained. And the two of them sat waiting, as in a wake for the child. The woman seemed to have accepted what she did not comprehend. She no longer prodded the baby or nuzzled it. She sat with it lifeless in her arms and stared ahead.

"I am so sorry," Garcia said.

29

IF THE DICTIONARY USED A PHOTO TO ILLUSTRATE *firefighter*, the photo would be that of handsome Neal Chapman of the Miami Beach Fire Department. As easygoing and modest as he is good-looking—with his gray eyes, light hair, and chiseled features—the dashing fireman grinned shyly whenever his compatriots teased him about his Medal of Valor. Joking, the other firefighters visualized a mock Christmas play in which stocky Jim Barrett played Santa, Stu Merker an elf, and Neal Chapman Super Firefighter who flew to their rescue when they set their sleigh afire.

Like cops or jet pilots or professional athletes, firefighters are rough extroverts who often hide their more tender and noble emotions behind a barrage of rough banter and loud grab-assing.

"Chapman's never off duty," Merker decided playfully. "Find that kid a phone booth—and there he goes: *the Last American Hero.*"

Cut through the rough surface, however, and what emerges are people who, like a family, harbor deep affection and respect for each other. During the Christmas season, when Firehouse 2 brimmed with cheer and visitors, Merker or Mogen or Daugherty or maybe Creel gathered listeners beneath the Santa-bearded Hog's Head Chief and related the drama of Chapman's heroism. Chapman would never have told it himself. It was as if the story defined the heart and soul of who and what a firefighter really was.

"There are not many people kids can look up to in this world anymore," remarked Captain Luis Garcia. "Kids can look up to a guy like Neal Chapman."

Chapman lived with his wife across the Seventy-ninth Street Causeway in the City of Miami, in a northwest neighborhood mostly built following World War II, when people still believed in the American Dream. One morning, after a dry shift at the firehouse —only two runs during the night, both of them garbage runs—Chapman changed out of uniform in the station locker room and headed home. The relief shift checked equipment and started housekeeping as Florida's morning sun brightened the world.

Chapman stopped at a 7-Eleven for milk and bread. As he neared home, turning onto N.W. 147th Street lined with bright little Florida bungalows, he spotted black-gray smoke seeping from the seams of one of the bungalows. Someone's American Dream was on fire.

For a firefighter, encountering a house fire like that without his equipment and apparatus was comparable to an unarmed cop blundering onto an armed robbery in progress. Nonetheless, the off-duty fireman wheeled his car to the front of the house. Reacting instinctively, he jumped out of the car and raced toward the smoke. This early in the morning, the neighborhood lay somnolent, not yet up and moving.

The only person about was the next door neighbor, who apparently smelled smoke and thrust his head out through the crack of his door just as a lean young man in blue jeans and T-shirt raced across the lawn of the burning house. Chapman's first thought was that someone might be trapped inside.

"Dial 911!" he shouted at the alarmed neighbor. "Do you know who lives here? Do you know if they're home?"

"They have an old woman . . . a housekeeper . . ."

That was all Chapman needed to know. A half-dozen long strides brought him to the front porch. Through the window he saw flames bending and curling inside. Smoke pushed out from around the door. Ignoring any thought of his own safety, the firefighter found the front door unlocked. He flung open the dragon's mouth. The house coughed phlegmatic dark wads of smoke at him. Chapman coughed back. He hesitated only a second. Without air, tools, and water lines, he felt like an unarmed Wyatt Earp arriving to do battle with the Clantons.

From inside, from deep within the heavy smoke, came a human cry. That was all the incentive Chapman required. Undaunted by his lack of weapons, knowing only that he was a firefighter above all else and that firefighters put out fires and rescued survivors, risking their own lives whenever necessary, Chapman plunged into the house's smoky bowels.

It was as if the house grabbed him and tried to evict him. Not so long out of fire school himself, his memories were all too fresh of when instructors required probies to enter the smokehouse and unmask. Most probies held their breath as long as they could while the black smoke attacked their eyeballs and nostrils. Finally, when they had to gulp, their heads went light on them. Some panicked the first time and rushed out of the smoke, gasping and sucking desperately for fresh air. Chapman recalled that panicky feeling in the pit of his own stomach when he realized he couldn't breathe, when what he sucked into his tortured lungs was hot, acid, and thick—like he was breathing lava.

Now, that same panic rose unbidden in his gorge. Rushing *into* a fire was an unnatural act. His eyes stung as though they'd been dipped in acid. Working to control his own fears, coughing like a TB patient, he steeled his nerves and, driven by the human cry he thought he heard from the inside rear of the house, dropped low to the floor to where the fresher air lay and crab-walked rapidly across what he took to be the living room.

Keep cool, he told himself, keep on track.

All but blind in the smoke, he followed fresher air into what he took to be a short hallway. Feeling along the walls to either side, he came to a closed door. Flames glowed bright above him through the thick smoke gathered at the ceiling. Fire was apparently working its way up the walls.

His confused senses, feeling trapped, screamed at him to keep rational.

Smoke inside the first bedroom was not as dense as in the living room. Blinking away tears, still squatting, he picked out the cowering form of an elderly woman next to a twin bed in the far corner of the room. Frozen in terror, she appeared too frightened to either move or scream. Her eyes settled on the crouching intruder like they had spotted a giant tarantula or crab.

"It's okay," Chapman called out reassuringly. "I'm here to get you out."

The old woman replied in rapid Spanish, then broke out coughing. Not understanding Spanish, with no time to start learning, the firefighter simply rushed forward and grabbed the woman. She blistered him with a fresh onslaught of Spanish and wrenched away from him when they reached the hallway. She lurched back toward the door in a shambling, half-crazed run.

Chapman caught up with her. Although he had heard of such things before, he had never seen it happen where people lost their senses and stampeded blindly about in a fire until they fell from smoke and heat exhaustion.

"I'm a fireman," Chapman explained quickly. "Understand? Fireman. I'm here to help you."

She flailed at Chapman, with her voice and with her hands. Panic, he thought. Pure panic. He felt a little of it himself. Patience burned thin by the increasing heat and smoke, he grabbed the frail creature and bodily hustled her through the front-room smoke and out the door into the fresh morning air. She coughed and gasped and sobbed, but still tried to fight her way free of the firefighter.

Neighbors had gathered on the front lawn by now.

"I'm a firefighter," Chapman explained again. "Take care of this woman. I'm going to see if I can get the fire under control."

Looking back on it, it was a foolish thing to do—dashing back into the flames. But fighting fires, saving lives *and* property—*that* was what firefighters did. Back inside the house, the smoke had thinned because of ventilation through the open front door.

Chapman kicked over a burning coffee table and stomped out the flames. He ripped burning curtains off the windows and stomped them out.

Fire glowed from overhead. Giddy from smoke inhalation, he was about to give up his efforts as futile when, suddenly, he looked over and saw the old woman charging wildly back into the fire. Her form in the smoke outlined itself briefly against the open door. Then, with an insane Spanish shout, she flung herself across the living room.

Chapman jumped over the coffee table and caught her. *Crazy old woman!* What could possibly be valuable enough in here to cause her to risk her life like this? He tucked the frail, struggling form underneath one arm and once again carried her out of the house. He'd have a neighbor sit on her this time. *He* would sit on her himself. There wasn't anything else he could do inside the house. The bungalow was a goner.

Like everyone who had lived for any time in South Florida, with its huge Spanish influence, Chapman understood a few Spanish words and phrases. He hadn't been listening to the old woman before. But now he did. Shouting, screeching, pointing toward the house, tears streaking her face, the lady bellowed in awful fear the same word again and again: *"Ninos! Ninos! Ninos!"*

That was what it was about. *That* was why she kept trying to return to the fire.

Babies! There were *children* inside the house.

Chapman cast a horrified glance at the house. It didn't take fire long to spread. Flames licked through the roof. He took a deep breath to clear his lungs and oxygenate them. He gestured hurriedly, assuring the old woman that he understood and that he, not she, would rescue the children. That seemed to satisfy and calm her. She stepped aside.

One more time, Chapman coached himself. It's just a fire. It's your job.

He thought he heard sirens in the distance. He couldn't wait.

"Los niños!" shrilled the old woman, a babysitter, as it turned out.

Once again Chapman plunged into the house, by now an inferno. He crab-walked across the living room and down the short hallway before he had to expel his lungs. He sucked in smoke and heat. He went light-headed. Supporting himself with one hand against the wall, he quickly found the first bedroom inside which the old woman had cowered.

He had a decision to make. He realized that physically his time was running out. At most he had time to search *one* room before he succumbed to smoke inhalation. The old woman had seemed to be alone in her room.

He made his way, stumbling and wheezing, down the hallway to a second door. He flung it open. Bathroom. *Keep a cool head, man. Keep focused.*

He stumbled, coughing, and sank to one knee, head spinning, just as he reached a third door. He shook his head savagely, like a reeling boxer brought to the ropes in the tenth round by too many right hooks. Smoke clawed at his throat. He felt as though his lungs were being turned inside out and seared on a grill.

As though from a great distance, the choked wailing of babies penetrated the smoke that fogged his brain. The children were alive! A shot of adrenaline revived him. He struggled to his feet. The crying guided him, motivated him. The babies' lives depended upon him.

He smashed through the closed door and found the air better beyond. Peering through the smoke and his tears, he stumbled about until he found a crib containing two writhing youngsters, both young enough that they still wore diapers. Their little eyes bulged with tears, pain, and terror. They beat themselves against the crib railings trying to reach their rescuer.

No time to comfort them. Chapman snatched a sheet and rolled the struggling little bodies into it. Grasping the bundle in his arms while the babies frantically resisted this latest in a series of assaults upon their young sensibilities, the weakening firefighter fought his way back through the smoke and flames.

He glimpsed the lighter outline of the open door ahead. He stumbled and fell to his knees. He refused to stay down. Raw adrenaline and survival instinct drove him the final few feet to safety.

He staggered out the front door into the arms of another Miami Beach firefighter, Don Farrington, who also happened to live in the neighborhood and who, seeing the smoke, had just arrived. Chapman refused to relinquish the squirming children clutched tightly in his arms. Farrington eased the exhausted firefighter and his babies out onto the lawn away from the burning house.

Minutes later, after Miami firefighters were on the scene and Fire Rescue had whisked the survivors off to the hospital, Chapman caught his breath with a few whiffs of bottled oxygen and checked on the condition of the elderly woman and the babies.

"They took in a lot of smoke," he was told. "But thanks to you, they're going to be all right."

Humility is common where heroism is common.

"Just another day at the office," Neal Chapman mumbled shyly, striking the self-deprecating pose with which he would later accept his Medal of Valor. It was the perfect pose for his *firefighter* picture in the dictionary.

30

LIKE ALL FIRE STATIONS, STATION 2 UTILIZED THE TIME between alarms to maintain equipment and apparatus. Firefighters took a great deal of pride in the appearance of their trucks and pumpers. In earlier days, when fire departments were a city's social hub and the fire engine the center of any parade, firefighters began and nurtured the tradition of decorating their apparatus with fancy designs—roses, wreaths, elaborate scrolls, nicknames, and the like. Firehouses often competed with each other for the best-looking apparatus, the most elaborate designs. Lucky indeed was the firehouse that boasted its own artist. Firefighters painted in secret and then proudly displayed their trucks and engines on parade day.

While the tradition is fading nowadays, firefighters nonetheless would never be caught leaving the station house riding anything less than apparatus cleaned and buffed to its maximum gloss. Stu Merker once observed jokingly that the same firefighter who drove a rusted, mud-splattered Chevy to work would be insulted if seen on a pumper in the same condition.

Fire trucks as they are known today are handsome towers of strength, tough against their hot adversary, full of tradition. Fire trucks powered by internal combustion engines arrived in the United States in the early part of the twentieth century. They appeared as another step in man's eternal war against wild fire.

Man has lived with the force of fire since the beginning of time. Constantly, he searched for faster and more efficient ways of bringing it under control when it escaped. The ancient Egyptians constructed primitive pumps to fight fire, along with fire hoses made from the entrails of oxen. Romans formed fire-fighting brigades. Although history records these references to a variety of fire-fighting techniques, squirts and syringes and pumps and other equipment, it was not until the seventeenth century in Germany and England that the hand pump began replacing the "bucket brigade" which had constituted the universal method for fighting fire for hundreds of years.

Hand pumps evolved into pumps operated by steam mounted on carts drawn first by men, then by horses. In the late nineteenth century and early twentieth, manufacturers experimented with a variety of pumps—rotary gear, centrifugal, and piston. Whether the pump was pulled to the scene of a fire by men or, later, arrived in a self-propelled vehicle, it was the heart of the fire pumping engine. As early as 1917, Ahrens-Fox produced a piston pump that elevated a stream of water as high as the Woolworth Building in New York City. The modern pump is the centrifugal pump capable of pumping up to 2000 gallons per minute.

Along with pumps evolved hoses and nozzles. The standard hose, exclusive of the small "red" lines and booster lines, is made of rubber tubing covered by a woven jacket of double thickness. The jackets were formerly made of cotton. Today, cotton is being replaced by more space age polyester fiber and other "miracle" fabrics.

Three types of nozzles are in general use: the straight-stream type, most effective in outside attack on large fires, in three-fourths inch or 1¼ inch, capable of delivering 560 gallons per minute at seventy pounds per square inch; the spray or fog-pattern nozzle, for smaller fires demanding wide coverage; and the combination nozzle, the most popular and widely used, which can be set for either spray or straight-stream.

In 1907 Seagrave, a name still alive today, built the first self-propelled engine. Although horse-drawn carts remained around into the 1920s, their days were numbered. Frederick Seagrave, followed by Chris Ahrens, who designed the classic fire engine with a right-hand steering wheel, and then such enduring names as American LaFrance, Maxim, International, Reo, Ford, Willys, GMC, Dodge, Mack, Pirsch, and Cadillac, all entered the specialized field of fire engine manufacture. Some failed, some succeeded. Ahrens, for example, never recovered from the Great Depression.

While manufacturers constructed big rigs with ladders and capacity pumps for the larger cities, many also found a market in the smaller cities and rural towns in building fire apparatus that could be applied to stock vehicles. Some even offered kits adapted to virtually any old truck. One of the most popular vehicles for kit application was the Ford, beginning with the Model T and moving on along to the Model A and its V-8 configuration.

All of the older fire engines have not ended up in fire museums. Yet. Some are still working. In San Miguel, California, population six hundred, for example, the volunteer fire department drives a 1940 model Ford V-8. The siren is dented and the horn button cracked, but both still work. The tachometer doesn't, but it isn't needed anyhow. For maximum effectiveness the rusted, faded old rig, never repainted, is kept in the drive of the firefighter on call. It carries a thousand feet of 2½-inch hose and a tank of 250 gallons. It can deliver 500 gallons per minute.

Gib Buckman, whose father bought the truck for the city when Buckman was six years old, takes care of the old Ford and makes sure it is maintained.

"This rig gets me a lot of smart remarks," Buckman said. "Once, when I pulled out of my driveway, some guy asked me if I was going to a parade.

"'No,' I said, 'I'm going to a fire.'"

The arrival of streamlining in the mid-1930s made the enclosed cab a fashionable feature. In 1939 American LaFrance introduced the cab-forward concept in which the cab is set forward over the motor. Cab-forward ended the classical long hood and became a universal design still used today. The typical modern fire engine is cab forward with 2000 feet of hose and a centrifugal pump capable of delivering 1000 to 1500 gallons per minute, a 500-gallon tank, and two ground ladders of 35 feet and 28 feet.

When the alarm sounds at Station 2 on Miami Beach, the red Michigan Instruments steed that explodes from the double doors is both traditional and typical.

31

ORIGINALLY FORMED TO RESPOND TO TRUE EMERGENCIES, emergencies in which life and death hung balanced on a hair, Fire Rescue instead quickly transformed itself into the poor man's medical service. "We're the only game in town that makes house calls," Otto Ramirez quipped. Stomachaches, a cut knee, bruises, colds and sprained ankles, a call to maneuver Mr. Jones down from the penthouse or get a glass of water for the surviving alcoholic sister Hazel—these calls outnumbered the genuine emergencies to the point that it was almost a surprise to the paramedics when they ran hot and discovered they were *really* needed.

"I don't know how, but the guy's still alive and conscious," one of the cops at the Fisher Island construction site shouted at the arriving emergency van. "He looks really fucked up," he added in a softer aside after Jim Barrett jumped out of the van with his aid bag and Glenda Guise and Otto Ramirez in tow.

This was no chickenshit run. *This* was an emergency.

Running, the three paramedics followed the patrolman through an opening in the wooden fence that surrounded the high-rise construction site. They heard painful cries, sobbing. The cop explained the situation in a few words. A construction iron worker walking the high steel beams of the new high-rise skeleton lost his footing and plummeted from the fourth-floor level. He landed on a slab of concrete out of which protruded a number of sharp steel re-bar rods, each about two feet long.

The guy was pinned to the slab of concrete with three thin spikes piercing his body, like an insect pegged on display. Sweat suddenly rivuleted off Barrett's shaved head. His head gleamed as smooth as a wet light bulb. He hated it when tour rotations took him off fire suppression and landed him on Rescue. But this was a different call; he was needed on this call.

The worker was a young man with long brown hair. His body thrust up from the concrete on the steel rods like a sacrificial offering on an altar. Blood soaked his blue laborer's shirt where the points of the re-bar stuck up through his torso, tenting his shirt. Eyes blinking assured the paramedics that the victim still lived; he held his body rigid on the stakes against the obvious pain.

"Help me . . ." he croaked. His face twisted into a hideous grimace composed partly of pain and partly of raw fear.

Barrett slipped on the calm face he had learned to wear in front of victims. "That's why we're here, friend. Take it easy; we're going to get you to the hospital."

Swallowing their apprehensions, the paramedics conducted a quick physical assessment—blood pressure, pulse, breathing, bleeding. Vitals remained surprisingly strong: BP about 102/70; pulse rapid, 90, but strong; breathing labored but regular. That he was alive at all astonished the medics, who quickly got a lifeline IV into his vein to combat shock and counter loss of fluids.

Apparently the spikes had not penetrated vital organs, like the lungs or spleen. While Guise and Ramirez worked over the victim, Barrett dropped to his belly and peered into that space between the man's back and the concrete slab. The rods impaling his torso suspended him faceup off the concrete base by about three or four inches, while the free sharp ends of the rods protruded from the front of his torso.

One of the rods stuck up through the guy's chest between his nipple and right shoulder. Another had penetrated lower on the left chest area, sticking up bloody through his shirt from his rib cage. The third spike tented his loose blue shirt at the lower right abdomen.

Barrett's thoughts gridlocked for a few moments. He had to get the worker to the hospital emergency room. Every minute delayed detracted from his chances of survival. The only question was—how? How did you transport a man attached to a slab of concrete that looked to weigh about a ton?

Obviously they couldn't pull him off the rods like stripping a piece of cork off a football cleat. Even if they managed to avoid killing him outright by further damaging some vital organ, the shock of it would kill him. It would be more humane to take a cop's gun and put the poor bastard out of his misery.

Removing him from the spikes, therefore, offered no viable option. That left—what? Transporting the concrete too? How? Barrett glanced anxiously about. Cops and other construction workers watched him. He met Glenda's eyes.

How, her eyes questioned, are we going to get this guy to the emergency room?

She worked coolly over the victim, more coolly than Barrett felt, more coolly than perhaps she felt on the inside. She had the IV going. A normal saline drip replaced lost fluids and prevented arterial collapse and loss of BP. Ramirez cut off the victim's shirt and packed pressure bandages around the spikes and the wounds they made. He monitored and called out the victim's vital signs.

The worker bled, but not excessively. The skin puckered around the spikes, containing most of the blood loss to the body cavity. That put pressure on the heart and lungs. Pulmonary edema . . . Barrett ticked off the possible conditions and complications in his mind. Then he turned everything else off except how to solve their most pressing problem. Let Guise and Ramirez handle the emergency medicine.

Barrett had to get this guy transported.

In the back of that dump truck? It would take a crane to lift man and concrete into the truck bed. If he didn't die from that, the ride to the hospital would probably do him in. Riding the bed of a dump truck was comparable to driving a car with four flats down a rocky mountainside. A dump truck turned a healthy man's organs upside down.

"We're going to get you out of here," Barrett promised the victim.

Yeah? he demanded of himself. How are you going to do that, big mouth?

"Am . . . am I dying?" the guy asked, his eyes searching Barrett's.

"Hell no, buddy. What makes you think that? I don't want to hear it anymore, okay?"

If nothing else, Barrett would *will* life to remain. This was not some drunk wanting an aspirin against his hangover; this was a human being who truly needed rescuing. Barrett personally would not let him die.

Yet, he had to fight the feeling of helplessness sneaking up on him around the edges. His frantic eyes darted, searching. He had to think of *something*.

Maybe a chopper crane? Perhaps transport doctors and an emergency room team and their equipment to the scene?

He dismissed that thought. The guy wasn't going to last long enough.

"BP 100 over 68," Ramirez intoned. Blood pressure was falling. "Pulse 100." Pulse rising.

Not good.

"What happened?" Barrett asked the guy, to keep him talking and help ward off shock. It also provided Barrett a few more minutes to come up with a plan.

"I—I must have slipped or something—next thing I knew . . . I was playing Superman. . . ."

"What's your name, young man?"

"J-Joe. If—if I live over this, I'm never leaving the ground again."

"Okay, Joe. You're in good shape here."

Yeah?

"H-How are you going to . . . to get me off these things?"

That was the $64,000 question. Everyone—the cops, Guise and Ramirez, Joe's buddies—all looked to Barrett for the answer. Goddamnit, he was no miracle man.

Then, suddenly, the paramedic knew what he had to do. It was the only way. He summoned one of the policemen.

"Radio for the medicopter," he ordered. "I also need some padding—pillows, blankets, whatever. And a hacksaw."

While the cop rushed off to fulfill the odd request, Barrett continued talking to Joe in a calm voice. "Hang in there," he encouraged. And, "You're going to be okay, you hear?" Innocuous assurances that meant little unless you were on the receiving end. Barrett had seen before how such a simple thing as a confident voice in an emergency meant the difference between life and death. Something about it refocused a victim's desire to survive.

"I'm so—so damned glad you're here," Joe managed.

"So am I, Joe," Barrett responded. "So am I."

And so he was.

Becoming a firefighter proved a long haul for stocky Jim Barrett. His father had been a Miami smoke eater during the years before the Cuban influx. Some of Barrett's earliest memories were of his dad at the fire station wearing his big bumber boots turned out and over at the tops. Like little Manny and like the firefighters' children when they visited the station house at Christmas, Barrett as a young boy was held entranced by tales told of flashover fires and brave men laying lines and dogging ladders and risking their own lives to rescue trapped survivors.

By the time he was barely eight, firefighting had trapped him with its lure of action and adventure. He was a bullheaded kid with a chip on his shoulder and an urge to experience life. Impatient, he quit school at fifteen and took off on his own. At seventeen he enlisted in the U.S. Marine Corps. Although it was the tail end of the Vietnam War and he saw no combat, he thereafter credited the Marines with instilling in him the discipline to pursue his dream.

Fresh discharge in hand, he made the rounds of major fire departments in Florida. His disappointment grew as one after the other rejected him. The ex-Marine, personnel departments assured him, was simply too short of stature to qualify. The State of Florida had a requirement that all policemen and firemen had to be at least five feet eight inches tall. Never mind that young Barrett could pick up the rear of a Volkswagen, he was only five-six.

"If they measured him from shoulder to shoulder instead of from toe to head, he'd make it," quipped a friend at the time.

The Forest Service offered to hire him to fight fires—if he would get married. The job required a husband-wife team so the wife could man the watchtower whenever her husband was on a call. At the time, Barrett hadn't a single matrimonial prospect. Defeated there by his not having a wife, and elsewhere by his own lack of physical stature, he went to work for a Miami utility company. His dream lay in shreds at his feet.

The utility job bored him. In 1976, when Florida lowered its height requirements for firemen and policemen to five-six, Barrett was one of the first men in line for a fire job. He took a ten-thousand-dollar cut in salary to fight fires for the City of Pembroke Park. Four years later he transferred to Miami Beach. At Firehouse 2 he found a home.

Fire suppression was his passion; he thrived on the adrenaline rush of a fire run, sirening through the city streets while all eyes watched and all vehicles pulled over to let him through. Fighting fire, meeting it head on, made him feel important, needed. There were only a few professions that remained absolutely essential to modern society, without which civilization would bog down or go up in smoke; one of those professions was firefighter.

While Fire Rescue, he conceded, was also necessary, there were simply too many chickenshit calls for his liking. Times like now, when Rescue was truly needed, were too infrequent. On his hands and knees next to the impaled man on the concrete slab, speaking in soothing tones, Barrett had to admit that when the challenge came, however, it *arrived*.

The policeman returned promptly from his errand, accompanied by a construction worker who had ripped a pair of thin mattresses from the back of his good-time van. The cop handed Barrett a hacksaw he had scrounged from somewhere.

"Life flight's on its way," he announced. "ETA about seven minutes."

Joe's eyes bulged when he spotted the hacksaw. "Wha . . . ?"

Barrett explained. Paramedics dared not remove the spikes from the victim's torn body for fear of causing additional injuries; surgeons at Jackson Memorial would have to do that. But first Joe had to be freed from the concrete for the helicopter flight. Barrett proposed to saw through the rods between the victim's body and the concrete slab, leaving the re-bar in Joe until he arrived at the hospital.

It sounded easy enough.

Working carefully, sweat pouring from his smooth head, Barrett used one mattress to caulk the space between the injured worker's body and the concrete. That served several purposes. It prevented Joe's weight from driving him farther down onto the spikes. It stabilized his position while, hopefully, it absorbed at least some of the vibrations when Barrett started sawing.

It was a risk. It was a hell of a risk. The vibrations alone might kill the victim by tearing a major artery or further damaging an organ.

Barrett took up the saw. Joe's eyes followed him. Joe's eyes said he understood the risk.

"Do what you have to," he said bravely. His voice had grown noticeably weaker.

"We're getting you out of here," Barrett promised. He felt the sweat clammy in the palms of his hands.

Dropping to his belly, he carefully freed the first spike from its mattress padding. He took a deep breath, then concentrated on easing the saw into the limited space between Joe's body and the concrete. Metal saw teeth bit into the steel rod. Barrett felt the vibrations in his arm. Joe cried out in agony.

"Hold the top of the bar as steady as you can," Barrett directed Ramirez.

He brushed sweat from his eyes and continued sawing, slowly, smoothly, but continuously. Joe moaned softly. The saw ate its way quickly through the steel.

One down, two to go.

By the time Barrett started on the second rod, the victim, although still conscious, appeared too weak to emit anything other than an occasional soft groan. His eyes remained wide and trusting, fixed in near adulation upon his saviors.

It's a miracle he survived the fall, Barrett thought. It's a double miracle if he survives us.

Total silence engulfed the construction site.

"BP 96 over 60," Ramirez entoned. "Pulse 110."

Barrett's hands felt so slippery from sweat and blood that he had difficulty gripping the saw handle. He sawed through the second rod and moved to the last spike. Joe's body suffered a series of brief spasms. Suddenly he relaxed on the spikes that impaled him. Barrett shot a questioning look at his partners. Guise shook her head slightly, assuring Barrett that the guy hadn't died. He had fainted.

"BP 90 over 60," she said. "Pulse 120."

"Hold on, hold on, damnit," Barrett rasped at the

unconscious worker. He shifted to the third and last spike. He listened for the arrival of the helicopter as he worked.

"BP still dropping. Pulse rapid and weakening."

"Goddamnit, hold on."

The saw gnawed through the last spike. Barrett dropped his head onto his arms in relief just as a policeman gaping into the patch of sky between the high rises cried, "I hear the chopper." A moment later the white life bird filled the sky, dropping toward the street. It hovered before setting down.

The paramedics transferred Joe to the second mattress, then to a stretcher. They rushed him to the chopper and loaded him into its belly. Barrett and his crew stepped back, exhausted, as the bird clawed its way skyward and nosed over in the direction of the hospital five minutes away. Concrete dust, blood, and metal shavings caked Barrett's blue jumpsuit. Sweat stained it black underneath the arms and down the back.

With a glance at the now-empty sky, the construction workers returned to climbing the high steel, each careful to wear his safety harness this time. Policemen looked around. "It's all over, folks," they said to spectators. They climbed into their patrol cars and left. Onlookers muttered to each other and stared at the paramedics.

From the van, monitoring the radio, Guise shouted, "Okay, guys. Coffee break's over. We got another call. Somebody fell off a pier at the beach."

Barrett and Ramirez gathered their bags. As they hurried away, a voice from the high steel beams checked them.

"Rescue? You guys off to save somebody else's life?"

Barrett looked around.

"You are *all right,* man," the construction worker said. "We owe you one."

The paramedics grinned back at the guy. This one rescue, this single call when they were *really* needed, made all the chickenshit runs bearable. Barrett's spirits remained high for the rest of the day.

It was after dark before Rescue 2 went downtime and Rescue 2-2 took over first out. Barrett telephoned the hospital and learned that Joe would survive. He stepped outside the fire station into the dark for a leisurely cigarette, his first since the construction accident. When he returned inside, he said, "Some days I think the best thing I can give my wife for Christmas is to get some other job."

He shook his head, grinned. "But today ain't one of them," he said.

32

LITTLE MANNY WITH HIS BROWN TEDDY BEAR EYES AND coal miner's face, Manny the semi-orphan, used to stare when Ed Delfaverro strode across the common between the station house and the main administration building. As a member of the Fire SWAT team, the tall firefighter's appearance took on an additional sternness because of the heavy .38 caliber revolver sometimes strapped to his belt. To Manny, Delfaverro presented an anomaly.

The other firefighters arrived for work armed with bumber boots and wisecracks. By nature, Delfaverro possessed a quiet nature that sometimes made him appear stern and hid his gentle sense of humor. What Manny saw was a tall, serious man who came to work wearing a weapon which seemed to have robbed him of any normal firefighter levity. It was as if the thought of being placed in a situation where he might have to use his gun was simply so grave that Delfaverro had lost his ability to laugh.

"Is Lieutenant Delfaverro a fireman or a cop?" Manny asked once.

"He's a firefighter," Neal Chapman replied.

"Then why does he carry a gun?"

"He's a special kind of firefighter. He's like a doctor who goes with the police when there's trouble, in case someone gets hurt."

It didn't make sense to Manny. He hadn't known many doctors, but the one or two he met didn't carry guns.

"Has Lieutenant Delfaverro ever kilt anybody with his gun?"

These were questions Manny never dared ask Delfaverro.

"No."

"Will he ever have to?"

Chapman shot the basketball and bounced it off a rim. He stood with Manny and looked at Delfaverro walking away. "I'm sure he's asked himself that same question," he said.

Delfaverro had. Many times. Every Christmas, people who lived in a house on Pine Tree Drive down from Firehouse 2 placed a large cedar-bough wreath on their front lawn. Passing the wreath with its universal message of love—PEACE ON EARTH, GOOD WILL TOWARD MEN—invariably gave Delfaverro pause. The irony of it. Nations and men armed themselves to the teeth against each other while at the same time espousing Peace on Earth.

It was like the old Crusader's account of the siege of Hagia Sophia: "... And blood ran knee deep in the Glory of God."

"The only reason you carry a gun," tough police instructors preached when Delfaverro attended SWAT training, "is to use it if it becomes necessary. Think about that now, while you have time to think about it. Deadly force must never be taken lightly. Deadly force means you contemplate taking the life of a fellow human being. There is no longer a fleeing felon rule in Florida; you don't shoot some burglar running from the police. The single time anyone, including policemen, is authorized to use deadly force is in self-defense or to defend another from death or serious bodily harm.

"There is a delicate balance, then, between when you should use a weapon and when you cannot. You must be always reluctant to use it—but, at the same time, you must not hesitate to kill when it is necessary to kill. Hesitation at the wrong time means you could lose your life or cause some innocent person to lose his."

The tough police instructors paused dramatically. "Think about that delicate balance," they encouraged.

Ed Delfaverro had his first experience with "delicate balance" about a year after he completed SWAT school. Afterward he sometimes experienced night sweats. He snapped up in bed, sweating, while his mind replayed the sound of sirens shrilling in the distance, the cracking of gunshots nearer, and the way bricks and bottles and old carburetors and whatever else thudded around him like shrapnel from exploding howitzer shells. Fires glowed deep inside his brain, like they did that night he fled the howling mobs in the section of Miami known as Liberty City where arson fires silhouetted rioters on Sixty-second Street. Invisible strings from a giant puppeteer aroused the rioters against the backdrop of flames like characters from Dante's *Inferno*.

"Hell yes I was scared!" Delfaverro confessed later. "Only a *fool* wouldn't have been."

After Christmas 1988 and after New Years 1989 were winding down, when Ed Delfaverro was still considered a SWAT rookie, there came to Miami what started off to be the Super Bowl of riots. A Miami policeman named William Lozano, a Hispanic, blasted two black men off their motorcycle during a high-speed vehicle chase in Miami's black Overtown district. The incident might have gone relatively unnoticed as simply another entry on the police blotter, followed by routine police shooting board and state attorney's investigations, except for the bitter social and economic rivalry between Hispanics and blacks in South Florida.

Although the immigration of Haitians and other Caribbean peoples into Florida has greatly increased the black population, Hispanics retain status as the largest single population group in the greater Miami area, making up roughly half the population. The killing of black men by a Latin cop inflamed the animosity between the two races. January found black organizations leading protests against the shooting, demanding that Lozano be tried for murder. It was Florida's prelude to the Rodney King incident, which, a couple of years later, led to the Los Angeles rioting.

Inevitably, protests in the streets regressed to looting, arson, and random violence in Miami's Liberty City and in the Overtown district, both ghetto seedbeds of poverty. Police who had had years of practice at countering this sort of thing immediately sealed off the trouble zones, set up command posts, and fought in fierce little campaigns of territorial acquisition to retake the city. City services—police, fire, utilities—in surrounding suburban cities immediately went on alert to support Miami.

As in all the Miami riots since 1968, the City of Miami Beach placed its fire and police in reserve. Only a few bridges—called *causeways*—separated Miami Beach from the rioting. The firefighters at Station 2, as at the other three firehouses, were called in and placed at one hundred percent alert, ready to siren to the aid of their mainland counterparts. In Miami, every available cop and firefighter in the city, including SWAT teams and Fire SWAT medics, were hurled into the riot zones to work twenty-four-hour shifts, taking time off during lulls to snatch an hour or two of sleep.

"Overtime pay from the last Invitational built my home pool," quipped a Miami SWAT medic who had grown cynical with the regular recurrence of violence. "I expect this one to build me an addition to the house."

Medical personnel were always in short demand, especially medics trained to accompany police into the combat zone and work with them there. Miami Beach dispatched Lieutenant Ed Delfaverro and its four other Fire SWAT paramedics to Miami and placed them under the temporary command of Miami authorities. Delfaverro found himself operating with a four-man squad of Miami SWAT medics in Liberty

City, where the rioting had already deteriorated to gunfights between cops and rioters.

"Big Ed," joked Stu Merker with his impish humor, "is the first to receive his engraved invitation to the Invitationals. May God go with him into that dark night."

33

LIBERTY CITY WAS A WAR ZONE. THAT WAS THE ONLY WAY Delfaverro knew to describe it. It was his first riot. Astounded and shocked that this could be happening in an American city, he stood at the perimeter established by the Miami police and gazed into the raging heart of Liberty City. A bloodred setting sun, so big that it filled the end of N.W. Sixty-second Street, Liberty City's main thoroughfare, sketch-outlined running mobs and the eerie holocaust glow of burning buildings. The Miami Beach firefighter heard the spiteful supersonic-like crack of firearms banging. Mobs surged and roared. It was a primitive scene, savage, almost like an outtake from an old movie about the end of the world.

"Wait until it really gets dark," commented the policeman in riot gear standing next to Delfaverro. "That's when they all come out to feed."

As darkness rapidly approached, patrol cars rushed increasing numbers of casualties out of the combat zone to command post aid stations and ambulances. Ambulances refused to enter the riot. Most of the casualties were young black men with broken bones and lacerations, but rumors circulated that rioters, police, and even firefighters were suffering gunshots.

Word came down. The action had picked up. Miami police organized pushes into the combat zone to disperse looters ransacking groceries, pawnshops, and liquor stores. Liquor stores were always hardest hit. Each push launched from different sites around the perimeter consisted of from twelve to fifteen policemen supported by Fire SWAT medics. Immediate attention to gunshot wounds or other serious injury often meant the difference between life and death.

A young cop with his helmet face shield pulled down watched Delfaverro check his medical aid bag. The distant fires reflected in his face shield, smudging out the widened eyes behind it. Delfaverro glanced up.

"I'm glad you're going in with us," the cop said.

Other than dispersing the shifting mobs and eliminating looting and arson, police hoped to saturate the riot zone prophylactically—put so many cops on the streets that potential rioters either stayed in their homes or went to jail. A police patrol resembled a Greek troop phalanx with the officers garbed out in helmets, riot shields, and bullet-proof vests or flak jackets. As Delfaverro's patrol prepared to launch itself north on Thirteenth Avenue toward the main rioting where Sixty-second intersected, the cops nervously rechecked their weapons and gear and peered anxiously into the riot zone. Macholike, covering up their anxiety, the cops joked with each other.

"Verily, though I walk through the Valley of the Shadow of Death," murmured one, grinning tightly, "I shall fear no evil—"

"—for I'm the baddest motherfucker in the Valley," a second finished for him.

It was an old Vietnam saying.

Delfaverro and three fire medics from the Miami Fire Department accompanied the patrol. It crossed over the perimeter and advanced cautiously into a street that opened dark like a pathway through a midnight forest. Rioters had knocked out most of the streetlights during the day in order to have darkness after sundown for their looting forays. Although fire blazed and danced along distant Sixty-second Street, it cast little light this far away.

Following in the protected V of the police formation, carrying their laden aid bags and sidearms, the medics watched as a pair of police helicopters swooped low over a housing project, beaming down powerful floodlights to expose rioters like startled deer caught in headlights. Women with shopping carts filled with TVs, VCRs, and other loot scurried for cover. They resembled shoppers on Bargain Day at Wal-Mart.

A small boy of about ten or eleven darted out of the project and jabbed his middle finger at the passing police patrol.

"Mo'fucker!" he shouted. "Take you white asses outta here, mo'fuckers."

His mother caught him and dragged him off by the arm, scolding, "Shut up you mouf, Kolenole. I gone beat you ass."

Something about the incident prompted Delfaverro to reflect briefly upon the fact that while other mothers' sons were out stealing and looting and burning—shoulder to shoulder with their mothers in many instances—this one mother at least still attempted to fulfill her duties. Maybe she was failing, there were so many things against her, but at least she still tried. So many others simply gave up—and society was going to hell because of it.

A gunshot rang out in the night!

Another.

Policemen instinctively crouched with their firearms ready. Delfaverro had missed Vietnam; this was the first time he had experienced gunshots fired in anger. Perhaps gunshots fired at *him*. Like the policemen, the firefighter crouched in the street when the first shot rang out. Little knots of people concealed in doorways and alleys and other places in the darkness jeered and laughed unseen their scorn and contempt for authority. Delfaverro crouched in the riot-littered street, in the darkness filled with its mockery, and anger welled inside him and filled that part of him not already occupied by fear. He broke into a cold sweat that trickled down his face from underneath his riot helmet.

The policeman next to him carried a pump shotgun.

"I could kill every one of these thieving bastards and not blink an eye," he murmured. He paused. "We either kill them or one day they'll kill us."

Delfaverro blinked. The rage drained from him. It was confusing, this war on the streets of American cities; he resented it and he hated it already, and he understood somewhere deep inside that America had splintered itself into so many divergent and opposing minorities, all clamoring for more than their individual share of the American Dream, that the country was banging itself apart like a worn-out piston engine robbed of oil.

After a while, when no further gunshots followed the first two, the patrol silently moved on along the avenue, approaching Sixty-second Street, to which most of the more violent action had confined itself. Light from the scattered fires—torched cars and smoldering buildings—brightened until Delfaverro saw it reflected red and angry in the face shields of his companions.

They passed a new Chevrolet rolled onto its side in the street. All its windows were shattered, so that the patrol's feet crunched on broken glass. Beyond the Chevy rested more auto hulks, abandoned over the years and simply left where they stopped running. Some of them, scorched by fire, had been pushed into the street to block it. The patrol wended its way through the rusted and blackened tombstones.

"It's like Hue during the Tet Offensive," a Vietnam vet decided, his voice reflecting the tension they all felt. "This is it. I've had it. I'm retiring after this shit's over."

They came to a stop sign. Scribbled on the sign, like a tribal boundary warning, loomed the message: PIGS MUST DIE.

"Lovely," Delfaverro whispered.

Ahead, at the Sixty-second Street intersection, a noisy mob clogged the crossroad. At first there were just kinetic shadows flitting against firelight. A dog inside a fence barked passionately at the passing patrol. The shadows gradually materialized into individuals as the patrol warily approached, shields forming a wall. Instead of dispersing, melting away to form elsewhere, the rioters held their ground.

It was like two rival gangs facing off.

Delfaverro's tongue stuck to the roof of his mouth. It felt swollen and dry, like a fat lizard dried out in the desert sun. He stared, disbelieving, stunned by the audacity of this bunch. There were about fifty of them, he estimated, black men and boys armed with ball bats and butcher knives. He thought he saw two or three pistols, and he was sure the fat man to one side had a shotgun. They apparently intended to defend their right to continue looting without police interference.

Delfaverro felt heavily outnumbered. The palm of his hand rested on the butt of his revolver. Funny, he reflected later, how he, a paramedic trained to save lives, reached first for that split side of his profession trained to take lives.

"Disperse!" the police commander thundered through his bullhorn as the phalanx drew near. "This is an illegal gathering. You will be arrested unless you disperse."

The police halted. From that point, the confrontation advanced like an example from a SWAT textbook on riot control. The standoff during which the police attempted to break up the mob and scatter it— "Disperse! This is your last warning!"—drew a jeering barrage of epithets hurled like rotten bits of garbage. The looters pranced defiantly about, grabbing their balls and slapping palms and brandishing their weapons.

"Walk right up here and make us, you got the balls."

Next came the hurling of rocks, bottles, rubber balls spiked with razor blades, nail-studded golf balls, auto parts—raining down on the policemen like shrapnel. Police responded with tear gas. Snatching his mask, Delfaverro broke for cover with the other paramedics. It wasn't their job to stand up to the mobs; they patched up the wounded afterward.

Clouds of gas choked the street. In the darkness and the clouds of gas as he huddled behind the row of abandoned and stripped autos, Delfaverro followed the advance of the melee around him by the sounds of screaming, pounding feet, shouting, and then gunshots. Unlike cops, who are unaccustomed to visibility restrictions, the firefighter felt relatively at ease in the gas clouds. After all, he had penetrated many fires where smoke cut visibility to less than the length of his helmet visor. The thing that bothered him was that while fire held no personal animosity toward the warriors who confronted it—when it killed, it killed without malice or vengeance or glee—this different type of enemy he now faced breathed and shouted and *hated* with a personal fervor that chilled him literally to his spine.

He thought he much preferred fighting flames.

He felt almost naked when the gas thinned and then blew away on the tropical breezes. As he stood up and looked around, the first thing he noticed was that the police were gone, chasing rioters. The second thing he noticed, chillingly, was that not all the mob had vanished into the shadows. A threatening voice boomed out of the darkness between a line of row houses and a tenement complex.

"Looky them mo'fuckers. Get they white asses."

People poured into the street and charged toward the four isolated fire medics, hurling everything at them except the roofs off buildings. For that one terrifying instant, tall Ed Delfaverro had never felt so alone and vulnerable in his life. Those not in peril might safely talk "disadvantaged" and "disenfranchised" and how a "slave legacy" had created situations like this, they might talk it academically until they lost their breath—but it meant nothing to someone out in the streets about to be trampled and mauled by a mob that wanted his blood. One of the Miami Fire SWAT medics was black; he took off, fleeing first along the row of old cars while missiles clanged and banged around him. Delfaverro and the others followed.

The last car in line was an old Cadillac rusting on its axles. From there, south, the street lay wide and open without available cover. Although riot snipers weren't known to be accurate shots, Delfaverro hesitated to expose himself to let them practice on him. Desperately, the medics hurled glances in a dozen different directions looking for a way out. The shrieking mob bore down on them.

Wildly, it reminded Delfaverro of some old John Wayne or Randolph Scott Western. "Yep, pilgrim. The Indians has got us surrounded—but we're gonna take some of 'em with us. Save the last bullet for yourself."

Trained first to protect and save lives, Lieutenant Ed Delfaverro reached for his weapon. He felt the smooth grips in his fist; his finger curled around the trigger. Some of them, goddamn 'em, were going down first. If it was his last stand, he wasn't going out alone.

The thought shocked him. He realized at that moment that he could kill, that he *would* kill, if he had to.

"Keep your heads down and zigzag," he overheard one of the Miami medics instruct. "Don't blast the bastards unless they catch up with you. Now, *run.*"

Later, much later, the lieutenant laughed about it when Merker or Mogen or somebody kidded him about it. "You really think *anything* could have caught you, Ed?"

His exit from the Liberty City riots, he admitted, was indeed a most ignoble one, but nothing to laugh about at the time. The medics burst from the line of abandoned wrecks like a covey of frightened quail. The quail turned into foxes once they hit the open street, foxes pursued by a pack of baying hounds. It almost reminded Delfaverro, who liked old movies, of the big final scenes from *Frankenstein* in which the mobs pursue the monster, intent on destroying him. Casting anxious glances back over his shoulder, Delfaverro stumbled and sprawled onto all fours in the street. He scrambled to his feet. A gunshot echoed through the streets behind him. He turned his legs into overdrive.

Sweating and panting, trembling from excitement, he reached the police perimeter and the command post before he realized he still clutched his service revolver, still prepared to use it. He clutched it so tightly his fingers ached. He stared at the heavy weapon for a moment, as though not recognizing it or the hand that gripped it as his own.

How close he had come to using that gun! When might there be a next time?

He slowly holstered it.

PEACE ON EARTH, blazed the message from the large cedar bough on Pine Tree, GOOD WILL TOWARD MEN.

34

ONE MORNING THE ENGINE FROM STATION 2 SAT PARKED AT the curb before a row of beach bungalows. Merker, Creel, and a couple of other guys from the station stood around a thorn tree laughing and joking and poking each other and looking up into the tree's spiked branches. Some neighbors down the block came out to check on the excitement. Merker politely took a little lady aside, casting back mischievous glances that said such a prim little woman should not be fouled by association with a rowdy crew of firefighters.

The diminutive old woman wore her white hair bunned at the nape of her neck.

"I'm good to Puffy," she lamented, wiping tears, as Otto Ramirez and Tim Daugherty drove up in the Rescue 2 van.

"Cats are ingrates," Merker replied. "They like to test you."

Ramirez climbed out of the van and sauntered over to join the others. High above their heads in the tree, about forty feet up perched on a thin limb protected by vicious thorns, sat a black and white tabby. It meowed as it regarded its would-be rescuers with subtle contempt. It began tongue-washing its pads.

Ramirez grinned. It was classic firefighter/cat-in-a-tree stuff. It was almost a cliché. Over the years, Ramirez had answered his share of them—along with a parrot-in-a-tree call once. He had even rolled on a monkey-in-a-tree. Where else would a monkey be? The pet owners always appeared about to suffer a coronary while the victims, like this cat now, seemed quite content to leave circumstances as they were.

"Nurse Otto, how do you get a pussy out of a thorn tree?" Creel bantered. Merker had the cat's owner out of earshot, calming her.

"This happen to you often, Creel?" Ramirez shot back. "They spot you and run up a tree?"

"Treein' 'em is the only way ol' John could ever catch a pussy," the bantering continued.

"We made a run on it late last night," Creel explained to Ramirez. "I told the lady that if it was still in the tree today, we'd come get it down. It's still in the tree. Who's gonna skinny up those thorns?"

"It has to come down sometime," Ramirez pointed out. "Did you ever see a cat skeleton in a tree?"

"Miss Jessica ain't having none of that 'wait 'em out' philosophy. She wants Puffy down *right now.*"

Merker strolled over, thumbs hooked in his fire-red suspenders. Like he had the answer.

"There's a pole on the rig," he said. "Somebody go ask little Miss Marple for a glass of water or to use the telephone or something. Get her inside so she doesn't see us playing pussy piñata."

Dave Mogen shrugged. He ambled over to the pet owner. A moment later they entered the old lady's house together. Mogen turned and winked before the door closed behind him.

As soon as the elderly woman was out of sight, Merker produced his long pole. "I keep it for occasions like this," he explained.

It took two men to steady the pole and manipulate its point up through the tree limbs. The pole prodded toward the cat. The cat arched its back and slapped at the pole.

"Here, kitty-kitty-kitty . . ." Merker intoned sotto voce. "You want to be"—the pole moved in—"petted?"

The pole nudged the cat.

"Heads up!"

The animal sailed out of the tree, claws and teeth unsheathed and its hair spiked like a porcupine's quills. It landed in the grass about twenty feet from the tree, shook itself once, and then streaked underneath a neighbor's parked car.

"Fellas, it's absolutely true," Merker exclaimed, amazed. "They *are* just like politicians and lawyers. They always land on their feet."

When Miss Marple and Mogen returned, the pole had disappeared and the firefighters were loading up to leave.

"I guess we made your cat nervous," Ramirez explained. "She's over there underneath that car."

"Oh, thank you. Thank you. You're such wonderful boys."

In the van Daugherty turned to Ramirez. "That must have been a forty- or fifty-foot leap. It was incredible. I didn't think the cat would make it."

"Christmas," Ramirez replied, "is the season of miracles."

35

Although Glenda Guise preferred working fire suppression over Fire Rescue, she always put a little extra effort into Rescue during the Christmas holidays.

"I know what it feels like to have Christmas coming up and not knowing if you're going to live to see it or not," she explained to probies when they complained about having to work the red vans. "You're lying there bleeding and in pain—and then you hear the Rescue sirens. Suddenly, you're reassured. Help is on its way. You know you're going to make it after all. That's the kind of effect Rescue has on its victims out there."

Glenda had been on both ends of the sirens—as the

rescuer and the rescued. A couple of years before, just before Christmas, she and her husband Gary mounted their Honda motorcycle and turned north on State 836, the crowded flyover that linked I-95 to the Florida Turnpike. It was one of those gorgeous Florida winter days when, especially if you were young and in love, you thought you wanted to live forever. The couple intended to ride up to Fort Lauderdale on the beach for lunch, maybe catch an art gallery and do some Christmas shopping, then head back. Both had shifts to catch at seven A.M. the next morning.

During their few years on the Miami Beach Fire Department, they had routinely taken their turns on Rescue and pulled twisted and bloodied bodies from the wreckage of the auto crashes they witnessed and worked. Yet, on that gorgeous morning, death and mayhem and tragedy had been shoved to the backs of their minds. They had no thoughts of such matters when, ahead of them in the fast, thick traffic, Glenda glimpsed brake lights flashing suddenly. The gray Chevrolet sedan in the lane adjacent to the motorcycle swerved unexpectedly and cut into the bike.

Through Glenda's mind flashed brief images of the many emergency calls she had run with Hoffman, Ramirez, or Creel to scenes exactly like the one starting to unfold now—with one important exception. *She* and *Gary* had not been personally involved in the others.

Pain seared through Glenda's leg as the gray Chevy clipped the motorcycle. Gary fought the bike's wobble valiantly, but then he lost control, and the next thing Glenda knew she was airborne, hurled through the air over her husband's head and over the handlebars. Instinctively, automatically, her brain ticked off life-saving measures—control bleeding, clear airway, shock treatment; IV, BP and pulse and O_2, MAST and backboard...

You can't save 'em all, she heard Barrett saying from somewhere far away, even as her body struck the pavement and bounced like a toy doll thrown from the window of a speeding car. She heard the screeching of brakes as cars tried to avoid hitting her.

And then—minutes later? hours?—she was hurting. And her foot lay twisted, with shards of bone protruding through bloody flesh. And she was reaching.

"Gary...? *Gary!*"

She heard him nearby struggling to breathe. She crawled toward him, muttering his name.

...but you can save some of 'em.

Above the rumble of traffic and the nearer, excited babble of the crowd gathered around her, she heard the distant siren of Fire Rescue. Help was on the way. Help. Relief so tangible that it was like a warm bath washing over her. Paramedics would save her Gary.

For the first time she knew what it felt like to be on the other end of the siren. It was a feeling like being *delivered.* She relaxed, lay back on the concrete.

Rescue was on the way.

That was her last conscious thought for a while. The next thing she knew she was in the back of a fast-moving Rescue van. A Metro-Dade paramedic bent over her stretcher and smiled reassuringly when he saw she was awake. She failed to recognize the face, but that did not daunt her. It was a firefighter paramedic's face. It could have just as easily belonged to Hoffman or Merker or Ramirez, who took it all so personally. Firefighters were a most special breed, all with different faces but of the same face. They had that look that said these guys *cared.*

"Gary...?"

"Your husband's in the other van. Don't worry. He's going to be fine. So are you."

She mumbled something in reply.

The paramedic bent near. "I'm sorry. What?"

"Merry Christmas," Glenda murmured.

"Merry Christmas back. Damn. They made us take the mistletoe out of the van."

Glenda Guise smiled softly, drifting with the pain medication.

"I hope you took it down in my husband's van too," she said.

36

NO ONE EVER ACCUSED A FIREFIGHTER OF BEING A GOURMET, not in eating the concoctions whipped up by amateur houskeepers around firehouses. Stocky Jim Barrett slid a big hand across his bald pate and eyed the plate of food placed on the table in front of him. Cherubic Steve Hoffman, who, when he grinned, resembled more than ever Ben Walton, served as maître d'. Merker, wearing the Hog's Head Chief's Santa's beard, elbowed a place for himself at the table.

"Call him anything you want, but don't call him late for supper," David Mogen commented. He wore his red elf's hat.

"We look like Snow White and the Seven Dwarfs," Merker observed. "Dopey, Sleezy, Goofy, Rufus, Cletus, and"—he grinned at Glenda Guise—"Snow White herself."

Barrett stared at his plate. He sniffed suspiciously. Up and down the table Guise, Garcia, Harris, and the others in the firehouse family stifled giggles.

"You're kidding me, right?" Barrett grumbled, glaring at Hoffman. "I thought you said you were making lunch."

"This *is* lunch," Hoffman insisted, still grinning.

Barrett poked his fork at the pile on his plate. He sniffed. "What *is* it?"

"A culinary delight known as meat loaf."

"So *this* is what happens when the stray dogs disappear?"

"I made it," Hoffman announced proudly.

"Does the Health Department know about it?" Barrett asked. He lifted a limp green bean. Guise collapsed with mirth. "You made it in the microwave, didn't you?"

"That's what the damn thing's for," Hoffman retorted.

The machine sat for days wrapped in a red ribbon until Hoffman summoned the courage to actually use it. It had been a Christmas gift from neighbors.

Barrett looked appalled. "A microwave is for reheating stuff or boiling water," he admonished. "It ain't for cooking."

"Why not?"

"It poisons whatever you cook. It sterilizes your grandchildren and their grandchildren. You have to be careful about toxins. It'll probably make us glow in the dark."

Firefighters were so conservative, someone once

remarked, that when George Bush ran for President, a poll showed sixty-three percent of the nation's firefighters still voted for Eisenhower.

Barrett examined the other troops to see how many might have already gone neon. They were happily stowing away food. Barrett played with his, sniffed it, looked around.

The firehouse alarm discharged suddenly with a shrill clanging that reverberated throughout the station. Barrett jumped up from the table and grinned with broad relief.

"Saved by the bell!" he bellowed. "Let's get out of here before this damned meat loaf explodes."

37

FIREFIGHTERS SAID IF YOU WEREN'T SPIRITUAL BEFORE YOU came on the job, the job probably made you spiritual. There were two times especially when it made you appreciate life. One of these times was when a firefighter died on the job and forced all the others to face the fact that it *could* happen to them.

Like the time Miami Beach firefighter Lenny Rubin fell in the basement during the hurricane fire and drowned. Word spread quickly. Off-duty firefighters were either telephoned or heard the news on radio or TV. Some of them brought beer to Rubin's firehouse to help his buddies relax and regain their breath. They put their arms around each other, connecting to the brotherhood of the job, and talked about the dead firefighter.

For the funeral, the firehouse draped its apparatus in black to lead the mourning procession. Departments all over Florida and even from other states sent firefighters in their dress blues. Everybody couldn't get into the church; there must have been six hundred firefighters at the services. Everyone remarked on the closeness.

"Each time the alert comes in and we leave the firehouse," said Captain Jim Reilly before he died, "we know something like that could happen to any of us. One of us might not be coming back. It can be a little sobering."

The other time when the job made you appreciate life, although it didn't hit quite as close, was when someone other than a firefighter faced death in the flames. Confronting death in a fire taught humility. It taught that life was not forever, that it hung by a fragile thread. Those who have fought for life, went an old combat saying, learned to appreciate it with a verve the uninitiated shall never know.

A firefighter went through anything to reach people trapped inside a fire. Whenever he made a grab—rescued a life—the high of the afterglow left him walking on air, a feeling that was indeed almost spiritual.

When Jim Barrett covered the hotel fire on Collins Avenue on the second alarm, he had a feeling from the pushout that it was a bad one. It was three A.M. to begin with, an hour when even on Miami Beach most tourists had retired to their hotel rooms.

Driving, Barrett careened Engine 2 sharply to the left, where a huge mural depicting palms and beach and a road appeared to lead Collins Avenue right into the coral wall upon which the mural was painted. More than one drunk tourist had, to his dismay and regret, followed the avenue into the painted scene. Collins Avenue opened up past the curve at the mural, sheening back the after-hours glitter of the dying tourist mecca. Hotels and condos separated Collins from the man-made beaches like tarnished stones studding an ancient necklace. Smoke rose from a four-story economy hotel at the far end of the necklace. White paint peeled off the face of the old hotel like cheap pancake makeup from the face of a dowager. Smoke seeping from it smudged out against the stars and moon.

Truckies from Station 3 had already arrived, along with a pumper and Fire Rescue. They were dogging aerial and tower ladders up to the old lady hotel's face even as the thick odor of wood and furniture smoke reached Barrett's nostrils and he saw for himself the reason for the frantic scurrying about. People, he thought, were about to die in this fire.

Hanging out windows, their eyes round and white like the big eyes from the old, once-popular Keane paintings, screaming and shouting and wailing, tenants clung to the face of the hotel like flies to a screen door. Barrett's horrified gaze fastened onto one young woman who hung to a window ledge by the tips of her fingers while her bare feet thrashed in midair fifty feet above the deadly pavement. Hot winds sucked into the fire, as into a bellows, flapped her nightgown like the wings of a stricken bird. Her shrieks tearing into the night, ripping it to shreds, mesmerized the arriving firefighter. He braked the pumper and stared up into the smoke as truckies scaled a ladder erected next to her. The truckies resembled ants. The girl seemed not to notice the ladder. Flames licking boldly out the window to whose precarious ledge she clung totally commanded her attention.

Barrett expected to see her drop smashing to the pavement within the next moment.

He hadn't realized he held his breath, that his lungs were aching for air, until the top truckie on the ladder grabbed the girl's thrashing body and pulled her toward the ladder. Hysterical from her imminent death, the girl struggled at first, afraid to let go her fingerhold. The truckie hooked his elbow around a ladder rung. Both firefighter and survivor swayed dangerously, but the truckie had snapped himself to the ladder. He held onto the terror-stricken girl, clutched her to him until she realized that she was being rescued. She sobbed pitifully then and clutched the truckie, like a small frightened child held onto its mother, as the firefighter carefully descended with her.

Barrett's lungs exploded and he sucked in fresh air.

As soon as the girl's feet touched ground, she ran away in pain and panic with her hands thrust out in front of her, zombielike. Someone from a Fire Rescue grabbed her. Cooked flesh dripped from her hands.

Most of the other survivors, while certainly in danger, were not in immediate peril. They waved and shouted with their heads stuck out their windows to gulp the fresher air. Barrett guessed blazes and thick smoke had cut them off from the hallways, stairwells, and elevators. Firefighters laid pipe to the hotel lobby, intent on knocking a pathway through the flames and clearing an internal passageway for rescue. Barrett guessed the effort would not be in time. The officer in charge stood on top of an engine cab with his radio and called for another alarm. The battle needed more ladders.

Already, on the fourth floor, a man wearing only

undershorts crawled from his smoking window and crouched on the windowsill. He unraveled a rope made of bed sheets knotted together and tied to a bed or something inside. Without checking the length of his makeshift rope, he dropped his weight onto it and started down hand over hand to the next lower window. The sheets proved too short. He hung on his line like a monkey on a rope and shrilled for help.

Truckies hustled victims to ambulances and Rescue; there weren't enough of them to man all the ladders.

"Barrett, Chapman, Merker—support the ladder company," the fire officer ordered through a megaphone. "Get those people down from there."

Tremendous heat blasted Barrett as, donning his SCBA, he dashed across the front of the inferno to relieve the truckie who had just rescued the girl from the ledge. The guy was sucking bad air; he had to change O_2 bottles and regain his strength. Stocky Jim Barrett started up the aerial ladder as soon as its crew shifted position toward the man dangling from the bed sheets.

"Reach out and grab the ladder," firefighters shouted up at the man.

"I can't. I can't let go!"

"We'll knock him off his rope if we try to get closer," a truckie yelled up to Barrett.

Barrett's concentration fixed on the man. The head of the tall ladder reaching deep into the billowing smoke rested within a foot of the survivor. Nevertheless, the guy continued to scream for help, apparently so terrified he couldn't turn loose the sheets and try for the ladder, not even with one hand.

Just as Barrett gained the second floor on his way up, the sheet itself caught fire.

Barrett sucked air. Gaze fixed on the burning sheet,

mesmerized by the implications, he scrambled up the ladder toward the doomed man. Smoke swirled around him. Even the air in his bottle heated up. Never before, and never since, had Barrett scaled a ladder so quickly. He went up the ladder like *he* was the monkey.

"Tarzan would have been proud of that Cheetah," Merker joked later.

When adrenaline pumped, men sometimes found themselves performing superhuman feats.

As though the sheets were a fuse, tendrils of flame edged down the cloth toward the survivor's hands, leaving black places behind, where more fire gnawed relentlessly into the material, weakening it.

"Grab the ladder!" Barrett roared. "To your left, goddamnit. Can't you see the goddamned ladder?"

The man hung on, frozen by fright. In another second or two he would plunge to his death.

Driven to recklessness, watching the awesome spectacle play itself out at the window above instead of keeping his eyes on the ladder rungs, Barrett miscalculated; his booted feet slipped. For one breath-caught instant he hung in midair by his hands.

Then he caught himself. Heart racing two steps ahead, he called upon every reserve ounce of speed and stamina he possessed. Above, still out of his reach, the victim dropped a heart-stopping inch or two as a flash of flame ate at a weak black spot in the street. Reaching, reaching, scrambling up the dizzying heights, Barrett felt as helpless as when he first looked upon Joe the construction worker impaled upon the steel re-bar spikes.

The rope of sheets stretched slowly at the burned places. Barrett was still ten feet below. Wide-eyed, the guy stared at the approaching firefighter. His eyes pleaded.

"Help me?" he called out weakly. His eyes cast a hopeless look below. Nearly four floors down. Nothing below through the smoke but concrete. And certain death.

"Help me?"

Barrett grabbed for the guy just as the sheets finally parted with a soft snap. For a moment that seemed frozen in eternity, the man hung suspended in air, as when a high hurdler going over the bar reaches his apogee.

It seemed the victim might slip on through the firefighter's grasp. A physically weaker man could never have pulled it off. More from desperation than design, Barrett held onto the sweat-slippery body. He jerked it toward him, wedged it in between himself and the ladder.

Sudden extra weight vibrated the ladder; it swayed dangerously, as though to toss both forms into space. Cold fear swept through Barrett as he realized he might plummet to earth. But he held onto the guy; he could not have released him even to save his own life. Firefighters never quit; firefighters always did their job.

An elbow hooked around a rung and Barrett's muscular arm saved them. And God, of course. Barrett always said it: God looked out for firefighters.

"Stop kicking and yelling!" he shouted.

His eyes made contact with the other man's.

"Calm down," he said. "Hold onto the ladder. You're safe."

"Safe?"

"Safe."

They descended together, hand over hand down through the choking smoke that surrounded them. The man disappeared as soon as his feet struck solid ground. Barrett looked around—and the guy was gone. Gone. No thank you. No kiss my ass. Nothing.

Barrett shrugged.

"Merry Christmas," he murmured.

38

JIM BARRETT HAD NO TIME TO REFLECT ON HIS RESCUE OF the man from the bed-sheet rope. He had that good feeling, that high feeling of accomplishment, but there was no time to pat himself on the back. His job wasn't over. Pumpers beat the flames back from the hotel lobby with their pressure streams of water. The fire officer ordered an internal assault.

"There are still people in there!" he yelled. "Haul your asses. Get them out of there."

All the visible survivors had been plucked from the windows and off the front of the old building. From the second floor up, flames, wind-whipped, blazed out of windows. Tongues of fire licked fifteen feet into the air above the flat roof. Trash rattled in the street from the backwind the fire sucked into its hot lungs. Heat threatened to spontaneously ignite even firefighters garbed for war in their fire-resistant turnouts.

"Anybody still alive in there, it's a miracle," opined an anonymous fire eater, his mask sooted with smoke.

Sometimes you had to believe in miracles. Barrett exchanged air bottles, gulped Gatorade, and joined his company inside the hotel lobby. Main lines and booster pipe crisscrossed the floor. Hard bolts of water knocked the flames back. Fire still roared from the upper floors, daring the firefighters to enter *those* caves to do battle.

Barrett joined Steve Hoffman and David Mogen. Wielding a two-inch line like a machine gun, they fought at the foot of the wide staircase, attempting to open a pathway to the rooms upstairs. By now no one expected to find anything upstairs except corpses. But they had to try.

Mogen, with Glenda Guise and John Creel, dragged a small backup hose to the staircase and opened water, while outside, an aerial attack flooded the roof and shot water into upper windows. Barrett and the Station 1 crew waded water ankle deep as it rivuleted down the stairs and streamed across the lobby floor and out the door.

The flames had a good hold on the hotel.

"Hey, Mogen," witty Stu Merker yelled. "Ever think of circumcision by fire?"

Mogen was doing a turn on a nozzle, blasting the wall of flames ahead. He mumbled a reply through his SCBA.

"You're getting weak, Dave," Merker shouted. "You need to take more vitamins."

Merker always had a comment. His comments kept up morale.

The internal assault gained ground on the enemy, then lost ground, beaten back repeatedly as the fire resurged. Crewmen rotated on the nozzles, each getting in nozzle time. Their muscles ached and their arms felt loose and strained in their sockets. Barrett had little concept of the passage of time until he rushed outside to exchange air bottles and down Gatorade. He breathed deeply and noticed the sky dawning pink above the Atlantic.

Another lovely day in Paradise.

Tall Vance Irik stood nearby in the street, studying the color, size, and location of the flames and smoke. He and his shovel would go to work in the rubble after the fire was defeated. Barrett grinned wanly. "Vance's Shovel." Sounded like a title for a children's book. But although abused children committed many arsons, arson was not child's play.

"Anybody else in there?" Irik wanted to know.

Barrett let his frustration show. "Damn it. We can't get out of the lobby. We're about to try again."

Finally, the firefighters cut a pathway to the staircase and climbed it gingerly. Merker and his crew battled into the first-floor hallway to search the rooms for unlikely survivors; Barrett led his patrol up the stairs to the second floor, where his powerful arms controlling the nozzle of the two-incher turned the hose onto fire that hissed like a furious dragon in its lair. Fierce flames slapped the water back. The three firefighters dragging the heavy line inched their way along the hallway through thick smoke that cut visibility to inches. Barrett tested each step with his weight as they advanced, feeling to make sure his footing remained solid and that the fire hadn't set a trap for them by eating away everything except a paper-shell veneer for them to fall through.

As they came to each door along the hallway, feeling for it, one of the three firemen darted into the room to check it for victims. Because of the smoke, they had to walk literally every inch of the rooms, including the bathrooms and closets. They peered underneath beds.

"Empty," they discovered, one after the other.

While the fire on the second floor generated copious

amounts of smoke, it appeared to have confined itself to the walls and ceilings and to other concealed construction chases. The walls were masonry, the floors wooden. Ahead in the hallway a firestorm of debris rained down. It looked like meteors flashing dimly through the smoke, as through a murky space.

"The building is about to go!" Barrett shouted. His orders commanded him to search the second floor and then get out of the building if it appeared in danger of caving in on them. Merker's crew searched the first floor; truckies had climbed ladders and entered the third and fourth floors.

"Looks like there's only a couple of other rooms to go," Mogen responded, his voice muffled through his mask.

Lieutenant Barrett made the decision: "We can't leave somebody behind."

After all, he noted wryly, God looked out for firefighters—and for fools.

The next room was unoccupied, like all the others before it. Pressing their luck, peppered by smoldering rubble, the rescuers fought their way to the last door along the hallway, an inside door. Hoffman took over the nozzle. He adjusted the control lever to Spray, drenching the hallway from ceiling to floor, wall to wall. Barrett tasted bad air at the bottom of his tank as he entered the last room and peered through the smoke like a chicken after sunset.

He found the bed. It hadn't been slept in. He dropped to the floor and looked underneath it. Just as he turned to get the hell out, something caught his attention. He moved toward it quickly and found the body of a fully-clothed man wedged between the bed and the wall.

"All right," he called out to his crew. "We got one in

here. Looks like he's a goner—but we got to get him out."

The victim was so large, obese, that the firefighter had to jerk the bed loose from where it was attached to the wall at its headboard in order to get the body out. Luckily, although fire gnawed at the ceiling above, it had not yet reached the corpse. The guy might have been drunk and lying in bed with his clothes on when the smoke got to him. He had rolled over, tried to get up perhaps, fell off the bed and found himself trapped between the bed and the wall.

Poor bastard.

Mogen ran in and helped hoist the man's dead weight into a fireman's carry on Barrett's broad back. Barrett staggered into the hallway beneath his burden and felt his way slowly toward the stairs, breathing heavily from exertion. Hoffman and Mogen fought a rearguard action with the hose.

Downstairs, the ground floor and the lobby had been saved, flames beaten back and out. Barrett considered the upper floors lost. About a dozen firefighters had moved into the cleared lobby to prepare for a blitz assault in force. Otto Ramirez working Rescue 2 met Barrett and helped the exhausted fighter lower his load to the floor. Barrett dragged himself to the wall nearest the door and collapsed, too spent to continue. He watched as Ramirez worked on the guy.

"Is he dead?" he managed finally, weakly.

Hey, you couldn't save everyone. He rescued the man on the bed sheets. One for two wasn't a bad night's work. You stayed in the major leagues with a .500 average.

"He *would* have been dead in another five minutes," Ramirez returned, sounding incredulous. "We've got a pulse. He's breathing. Somebody give me a hand here—get him out into the air. Somebody get Barrett out too."

Barrett could walk out on his own. A hitter with a 1.000 average—two for two, both homers—ran the bases himself. Figure it—a guy surviving that inferno upstairs for over an hour. Sometimes it was harder to kill a man than most people supposed.

Indeed it had been a good night's work. Barrett emerged onto broad Collins Avenue, now lighted by a predawn sun. Spectators knotted up behind police ropes. They applauded when the firefighters brought out the survivor. Merker might have taken a bow or something, but Jim Barrett simply walked into the street and removed his mask. The cooler air felt good on his fire-blackened skin. It tasted fine in his lungs.

He stood there away from the fire, watched the sun rise out of the ocean and felt the morning breezes. Nearby, Fire Investigator Vance Irik made notes on a pad. He glanced up and winked at Barrett. Barrett grinned back.

"Know who I saw at the hotel fire?" a firefighter commented later during cleanup at the station. "Remember Manny, our orphan? I think I saw the kid in the crowd. He looks all grown up."

39

RIDING RESCUE PROVIDED OTTO RAMIREZ WITH A SENSE OF the true value of human life. Life could be cheap. Cheap. People spent it so frivolously sometimes, squandered it on suicides, homicides, foolish alcohol- and drug-related accidents. Especially during the holidays, the Silly Season which seemed to produce something fundamental and dark in the human psyche. The Silly Season produced extremes. It produced great joy and great sorrow, warm memories and deep depressions.

So people squandered their own lives in one way or another. Ramirez accepted that stoically, philosophically. It was human nature. What he found difficult to accept was when another person, especially a person in authority or with special obligations, squandered someone else's life. For days after the accident at the mural wall on Collins Avenue, Ramirez went around cloaked in a deep sadness tinged with bitterness. Even when he had a few Christmas drinks—and he seldom drank—he couldn't rid himself of the feeling.

It didn't help matters that he was tired all the time anyhow. With Christmas approaching, Fire Rescue from all four Miami Beach firehouses frantically crisscrossed the island on emergency calls, literally meeting each other coming and going. Five or six of the department's eight emergency vans might be out on calls at the same time. Blue-eyed Ramirez rubbed his eyes until they reddened and swelled. The increased workload plus his extracurricular activities gave him the appearance of a walking wounded. Finals at nursing school required long hours of study, in between dealing with problems with the children and attending marriage counseling with his wife. It had been a tough year for Ramirez.

Even Mogen, with his red elf's hat and jokes about his pending ceremonial circumcision and Jim Barrett's poring over Christmas catalogues pursuing the perfect gift for his wife, failed to make the handsome Rescue man smile.

"Otto's got the nasties," Merker teased, trying to bring his friend out of it. "Santa Hog's Head Chief won't come down *his* chimney."

"I'd shoot his porker ass if he tried," Ramirez shot back, trying to come around.

The decaying odor of blood inside Rescue 2 seemed to reflect Ramirez's sour mood, one his friends hoped was temporary. Each new victim added to the stench. Shooting victims, stabbing victims, accident victims. Falls, drownings, beatings, rapes, muggings, ODs . . . God, it went on and on.

An auto crash victim with a severed femoral artery bled so much that blood sloshed around in the van like someone had tossed it in with a bucket. Ramirez threw open the van's back doors and washed it out with a hose. The blood had already seeped into seams and cracks and settled in the vehicle's pores. It rotted, and as it rotted it gave off a stench that settled in Ramirez's attitude. No matter how many times Ramirez hosed out the van, it still stank.

It smelled of death.

"Where you got the corpse stowed?" Daugherty or Guise or somebody asked.

"You're sitting on it."

There was no need to carry around a corpse. There was plenty of them around, harvested daily off the streets. If someone costumed and arranged them, there would be enough life-sized nativity scenes to keep the Supreme Court busy with church-and-state cases for the next fifty years.

Ramirez couldn't dislodge thoughts of that one guy from the mural and how and why he died. Hey, maybe the guy *was* a doper and a drunk. Perhaps his life wasn't worth much on the big scale. But it *was* a human life—and it *was* worth more, surely, than how it was spent.

The thing that tormented the paramedic was not the accident; he accepted accidents and their taking of lives. The thing was, somebody else was entrusted with the guy's life, and that trusted somebody else spent the guy's life. Ramirez knew it. He knew it as certainly as his van stank of death.

The victim was a passenger in a four-door Ford sedan with another young man and a driver who sped north on Collins and failed to negotiate the sharp left curve at the mural wall. The mural of palms and beach stood life-sized on its coral wall at the exact curve in the road, providing the southern length of the avenue with the illusion that Collins ended at the sea. Tonight's driver was not the first who, tanked up with too much beer or dope, had tried to drive into the idyllic scene.

The wall crumpled the car like a crushed beer can. The driver and the backseat passenger remained inside the car during the crash. It shook them around a bit, like pebbles in a tin can, but their injuries appeared relatively minor. They seemed more drunk or doped-up than injured, standing outside the wreck and staring in awe at the third guy and mumbling to themselves when Fire Rescue arrived.

It was this third guy who exploded through the windshield from the front passenger's seat upon impact and who now sprawled cut and bleeding and semiconscious on the hood of the car at the exact point that mangled metal met wall. Sirening toward the scene, Ramirez blinked, momentarily stunned, and took a second look. From a distance, with the mural light shining, the guy became a part of the scene itself, as though stretched out on the painted beach among the painted palms beneath the painted sunlight.

"Talk about tourist traps," a cop marveled.

Ramirez and Merker literally pried the victim out of the hard coral, leaving behind, indelibly imprinted upon the wall, his bloody silhouette. A quick physical assessment revealed broken bones and numerous lacerations, including a severe cut on the wrist from a shattered highball glass, the remains of which the man still clutched in a death's grip.

His heart tripped as erratically as that of a wounded wild rabbit's. Working quickly to stabilize before transporting, the paramedics nailed in a hasty IV lifeline to help maintain blood pressure and replace lost fluids. They immobilized suspected broken bones and, hoping to forestall shock due to trauma and loss of blood, outfitted the patient in a MAST suit—an inflatable body uniform that placed even pressure over a victim's entire body to preserve blood circulation and prevent the pooling of blood. Like many recent medical discoveries, MAST was developed for use on combat WIAs in Vietnam.

The guy remained very much alive and surviving when the sirening van reached the hospital in Miami that had one of the busiest emergency rooms and trauma centers in the United States. It was a major teaching hospital staffed by the finest doctors and the brightest nurses. Gunshot wounds were as common here as sprained ankles at other hospitals; accident injuries arrived one every few minutes. On Friday and Saturday nights especially, the emergency room resembled a combat field hospital.

"Pressure stabilized?" a nurse asked as the paramedics trundled their patient into E.R.

"Low but steady," Ramirez replied. "So far, so good."

"We'll have a doctor in right away."

Ramirez deposited the victim in a vacant room and stood outside the door, assembling his paperwork. A doctor came. He was young, but staff, not an intern. He nodded curtly at the paramedic as he entered the room. He hurried out again almost immediately. Curious as to why, Ramirez stepped inside, where to his dismay, he discovered the doctor had deflated the patient's MAST. Even the greenest intern, even a first semester nurse, knew that letting off pressure so rapidly might be extremely hazardous to a patient. Immediate E.R. treatment called for IV blood and a gradual deflation of the suit as the victim's vitals returned to normal.

Ramirez reinflated the suit. The patient moaned, still unconscious. As the medic turned away, he found the doctor standing at the door watching him.

"I deflated it for a purpose," the doctor snapped with that smug authority of the greater learned. "We're not concerned with his pressure."

"I am."

"We'll handle it from here, firefighter."

"Firefighter"—the way he used it sounded contemptuous, like a professor speaking to a first-year biology student who had grown a bit too precocious. Intimidated, knowing his job ended as soon as his patient reached E.R., Ramirez retreated to the door just as another, older doctor breezed into the room. The two doctors turned their backs on Ramirez, dismissing him. Boiling inside but holding his tongue, the paramedic stepped outside around the doorway to finish his reports. It wasn't that he was eavesdropping, but the doctors apparently failed to realize that he remained within hearing distance. Ramirez's ballpoint pen froze in its scribbling. He listened, both outraged and horrified.

"There is no apparent edema or other signs of internal bleeding," one doctor said.

"Are you certain? I was contemplating an exploratory inside as a teaching vehicle for some of the residents."

"There's no reason why we can't. All we need is a local. I'll get him prepped. You round up the residents and we'll provide them a little hands-on."

At first Ramirez refused to believe his own senses. But then as the disbelief wore off, he came to the only conclusion left him: these doctors were going to open up a guy for no sound medical reason other than to use him to teach other doctors. A living human being—nothing but a goddamned training aid.

The older doctor rushed out. He was tall and distinguished, with gray at the temples. He hesitated, obviously surprised that the Fire Rescue man remained outside the door and may have overheard.

"Doctor?" Ramirez began.

The doctor glared Ramirez up and down. His lip curled. "Yes?"

Ramirez might never forgive himself, but he stood there for a long moment, speechless, subdued. What could he say? That he was an undergraduate registered nurse, a fire paramedic with a few lifesaving courses under his belt? How did that stack up to this man's vast medical knowledge?

Besides, perhaps he had misunderstood. Perhaps these experienced medical doctors knew exactly what they were doing. He couldn't come right out and accuse them of something as bizarre and unethical as unnecessary surgery on a patient.

After a silent moment of confrontations, Ramirez backed down. He turned away, muttering, "Nothing. I must be hearing things."

"Of course," replied the doctor. "You did your job. Now let us do ours."

On the drive back across the causeway to Miami Beach in the predawn light, Stu Merker cast anxious glances at his silent partner. Something had happened to add to Ramirez's dispirited mood, but Merker didn't ask what.

"The guy'll make it. He wasn't that bad hurt," Merker said into Ramirez's blank stare, misunderstanding.

Ramirez stared out the window.

He learned the next day that the guy died during exploratory surgery.

Sometimes, he conceded, you couldn't save 'em all—even *after* you saved 'em.

40

"IT'S A JUNGLE OUT THERE," MOGEN TEASED, GRINNING, AS the big doors at Firehouse 2 rolled up. He stretched out his legs, relaxing in a good-natured, taunting way as Ramirez, Merker, and Tim Daugherty raced the alarm to their rescue van.

"You'll laugh out the other side of your ass when you're first-out tonight," Ramirez retorted.

Rescue 2 hit Pine Tree Drive. Somehow, lately, the cheerful colored lights glowing on Pinetree annoyed Ramirez. The annoyance stayed with him as he wheeled the boxy van, siren wailing, through the dinner hour traffic. The tourist crowds were gathering for another evening's entertainment in Paradise.

The call was to another auto crash, this one on Forty-first Street. It was Rescue 2's fourth car-accident run of the shift. It seemed Christmas revelers started out each day tanked up and playing bumper cars. Ramirez looked forward to the end of the Silly Season.

Sure enough, Rescue 2 rolled hot onto the scene—and it was about what Ramirez anticipated. More drunks, expecting someone else to pick up the pieces. They had wrapped their orange Firebird around a light post. The windshield shattered all over the pavement sent back sparkles from the streetlights, which had just flashed on against the gathering darkness. Cops waved their arms, blew their whistles and yelled at motorists who tried to slow down to gawk at the gore. A little knot of pedestrians, defying the cops, gathered around the Firebird to talk to the driver still seated in the wreck.

Blood spidered the driver's face. He was a young man who sat grinning goofily out his gaping windshield. Seated next to him, his wife, girlfriend, whatever, emitted a hefty barroom laugh. She wasn't bleeding, but her hair looked as though she had gone through a wind tunnel filled with crushed flying glass. In the rear seat huddled an older woman—the driver's mother, as it turned out. Her head lolled back against the seat, eyes rolling.

Carrying aid bags, the paramedics forced their way through the crowd. Sometimes, during his lighter moods, Ramirez felt there should be dramatic music playing when they arrived on the rescue. Like on TV.

The driver took a good look at the uniformed paramedics and grinned even more goofily. He exhausted a breath of eighty-proof air that knocked Daugherty back about three feet and snorted, "Well, all *right*. The cavalry has arrived."

His wife, girlfriend, whatever, cackled again and mumbled that she'd drink to that. She'd drink to anything.

Ramirez rolled his eyes.

"My arm hurts," slurred the older woman in the backseat, hitting the medics with wind off about six or ten ounces of good bourbon or scotch.

Mother, son, and girlfriend—all drunky as skunkies, as Merker put it, and feeling no pain. Anesthetized.

Ramirez and Daugherty got the young couple out of

the car, carefully checking for back and neck injuries, and stretched them out on the sidewalk underneath blankets. While they went to work on facial injuries and bruises, and ran vitals and started IVs against possible internal injuries, Merker took on the older woman. He left her in the backseat since she complained of back injuries as well as of her arm hurting.

In a sudden mood change, the driver bolted erect underneath his blanket and screamed, "Mom? *Mom?* Where's Mom? Is she okay?"

Ramirez consoled him, then glanced toward the mangled Firebird. Merker's head popped up from the backseat, mustache drooping, eyes wide, a caricature of himself. Ramirez had seldom seen the droll Merker without humor playing in his eyes and smile.

"Otto?" Merker said.

The way he said it—tense, a bit numbed—brought Ramirez to his feet. As a nursing student working on his degree, Ramirez was Station 2's medical authority. He left the young couple with Daugherty and hurried to Merker. Merker climbed from the car and drew Ramirez out of his patient's earshot. The older woman's moaning pursued them.

"My . . . my arm hurts."

"Otto, we need to get her moving ASAP."

"Broken?" Ramirez asked.

Procedure called for stabilizing a patient before embarking on the run for the hospital. A broken arm was normally not life threatening. Answering Ramirez's questioning look, Merker took a step, reached into the car and carefully lifted a large pad of loose gauze from his patient's head.

Ramirez blinked. A large section of the top of the woman's skull had been as neatly cleaved off as if a pathologist had used his little electric saw on it. Ramirez stared directly into the woman's brain. It was grayish and not that bloody. It seemed to pulse slightly beneath his gaze.

He thought she might be complaining about more than an injured arm if all the booze she'd consumed hadn't paralyzed her skull. As it was, when she regained her sensibilities, she was going to suffer a headache that only drunks and the worst hypochondriacs ever experience.

Merker thought they ought to find the missing part of her skull. The two paramedics scrambled around inside the car looking for it among the broken glass. They found a shoe sole, a torn book cover, and several beer cans, but nothing that resembled bone. Finally, giving up, they loaded the woman into the emergency van and transported her on a hot run. Rescue 2-2 on backup call transported the young couple more routinely.

After all the drunks reached the hospital, Ramirez filled out his paperwork and then, out of curiosity, checked on the old woman's condition before he left. She lay behind curtains in the E.R. while interns and nurses bustled about prepping her for surgery. A doctor came, looked at her exposed brain and pressed around inside her head with forceps, checking for glass and other debris. The booze must have worn off. She emitted a terrifying scream of pain that made Ramirez wince. He retreated quickly to the hallway.

She was about to have the granddaddy of all hangovers.

"How is she?" Merker asked as the paramedics strode to their van.

Ramirez almost grinned. "Sober," he said.

41

IF SOMETHING WENT DOWN ON MIAMI BEACH THAT REquired a fire investigator, and the dispatcher couldn't find Lieutenant Vance Irik anywhere else, she knew to beep him in state court over in Miami. Irik wasted so much time waiting around in the hallways of justice that he knew even some of the court stenographers by their first names. Justice moved slowly, especially in backlogged Dade County, where prosecutors pleabargained most burglaries, muggings, arson, and the like down to misdemeanors and a few weeks county jail time just to relieve the docket.

On a December morning when the sun refused to shine—one of those rare windswept days northerners came to Florida to escape—Irik hung around the Miami courthouse waiting to testify. He figured Rod Gibson, the defendant, would cop a plea at the last minute anyhow, before the trial by judge started, and for Irik it would be another day squandered. Witnesses waited, but they seldom testified. Maybe one in a hundred cases ever actually went to trial.

Weeks ago, Irik's expectations soared that the hotel arsonist plaguing Miami Beach might finally have run out of flame. An ancient transient hotel on the fringes of yuppie South Beach went up in flames. Word crackled over the radios that police had nabbed the firebug. Irik sped toward the scene, hoping.

When he arrived, he found Rod Gibson handcuffed in the back of a patrol car. The perpetrator resembled Charles Manson, especially in his dead fish's eyes and the offhand way he treated the matter. Police had nabbed him running wildly down Collins Avenue, chased by about twenty furious homosexuals from the hotel. Irik joined police in questioning the suspect. Being in Gibson's presence was like sitting in a deep freeze. He had been paroled from prison less than a month previously on an assault conviction.

"I gave them dick suckers fair warning," Gibson declared. He snorted a hard laugh. "Imagine that—being chased down by a bunch of pansies."

According to the arsonist, it was the gays' fault. In a flat monotone relieved only when he resorted to spats of anger, Gibson related how, a stranger in town, he inadvertently entered a seedy gay bar next door to the hotel to soak up a few suds. Being straight himself, he said, he noticed that a few of the other patrons appeared a little light in the trousers, but he thought nothing of it until one of them tried to hit on him.

"I'd have knocked his goddamned teeth out," Gibson declared. Only, some of the gays rooming at the hotel gathered around to taunt the stranger.

"You limp-wristed cocksuckers," Gibson called them. He left the bar, but not for long. When he returned an hour later, "I came back with a friend"—a beer bottle filled with gasoline and a chunk of rag stuffed into the opening for a wick. He checked the bar for his tormentors. Not finding them there, he looked into the hotel lobby next door, where he located them giggling and chatting.

"Like a bunch of teenager girls," he said.

"You have two minutes to get the hell out of here!" he shouted into the lobby. *"Then you're going to die!"*

He grinned at his interrogators later. "Maybe I didn't give them quite two minutes," he admitted.

In fact, the moment the alarmed gays scrambled like a covey of quail, Gibson lit his firebomb and hurled it like a fastball at the nearest sofa. Flames splashed. The hotel lobby ignited like a wad of dry paper. Gibson took off with the gays in hot pursuit.

"I'm glad the cops caught me first," he said.

Any expectations Irik held of clearing up his hotel arsons quickly faded. His pyromaniac still roamed Miami Beach.

"Man, that's the *only* hotel I torched. That's the only *anything* I ever burned. The cocksuckers deserved it. It's just too goddamned bad the fire department got there and put it out before it *really* scorched the butt fuckers' plumage."

Gibson laughed a hollow, amused sound. "It would have been worth going back to prison to roast a few peafowls."

The case ended up simply another arson case demanding Irik's presence in the halls of justice. He stood at the windows overlooking the sprawl of parking lots below. The morning remained gray, windswept. His beeper went off, startling him out of his reverie. The number that came up was that of the fire dispatcher. That usually meant trouble.

"Do what you have to do," the prosecutor told him. "We've offered your guy forty-five years behind bars, take it or leave it. He's an ex-con and he attacked gays. In a politically correct world, that's enough for a jury to get him life in the joint. He'll take the forty-five."

When Irik called dispatch, she recited a residential address on north Miami Beach between the Atlantic and one of the canals.

"Vance, we have another fire fatality," she said.

42

A POLICE CAR SAT PARKED WITH THE FIRE APPARATUS AT THE address of the small, decent house on north Miami Beach. It was a typical South Florida house, low and thick-walled against the tropical heat, with a red-tile roof. Fire pressure had busted out the front windows and turned them into ugly black eyes. Flames had gutted the interior. The house still smoldered.

Inside the police car huddled a chunky woman with ratted hair and a twisted, tearstained face. She stared out the car window, stared high past her house at the gray sky. She looked trapped. Something about her made Vance Irik think she had always looked trapped and might look trapped for the rest of her life.

"It's her house," a young firefighter from Station 3 explained, meeting Irik. "It's her baby inside the house."

"The fatality?" Irik responded automatically, not wanting to see it but knowing he must.

"Six years old, he is. Better get your shovel."

"I have to see it first. Let the police do their thing."

The sad sky dropped down like a worn cloak over the house and the firefighters. Irik sucked in a deep breath of the fresh air outside before he followed his guide. The living room was smoke- and water-stained, but otherwise relatively undamaged. The two bedrooms lay off a short hallway. They were gutted. Everything in them was black and gray ashes, everything melted and charred into everything else.

Including the six-year-old boy.

Irik stopped next to what remained of the bed. It had burned down to springs and metal. Wisps of smoke rose from it and from the little corpse. The corpse smelled a bit like steak on an outdoor barbecue. Crispy critters didn't stink much until you tried to move them and the crust busted open.

Just a baby, a little boy. His eyes had been burned out and his nose burned off, leaving tiny black holes in a face that resembled melted wax. Irik stared. Then he turned and slowly left the room.

Let the police do their thing, measuring and taking photos and collecting evidence. Irik and his shovel would wait until the fire heat left the house. Then he would reconstruct the scene in order to go to court and testify exactly as to how the fire occurred. Sifting around in the debris and ashes—where one piece of rubble represented a chair or nightstand or something, another represented what was left of the bed, with the hunk of burned flesh on it that had once been a little boy getting into mischief and playing ball with his pals.

Along with the scene, Irik would partially reconstruct the last hours of the lives of the people whose actions led up to one charred youngster lying in a burned bed.

A policeman summarized the statement from the trapped woman in the patrol car: "The way she tells it,

Vance, is she was pissed off at her boyfriend for not coming home last night or something. She piled a bunch of his clothing on his bed, poured kerosene over them—and struck a match. That'd teach the sonofabitch, eh? Burn down her own fucking house to get back at her boyfriend."

If you saw enough of these things, you believed people were capable of *anything*.

"Hey, when some people get pissed, they're *pissed*," the cop said.

"So how'd the little boy get involved?"

Irik's guide, the young firefighter, stood nearby listening. His face narrowed through its coating of grime and soot.

"Her son stayed overnight with a neighborhood buddy," the cop said. "She didn't realize when she set the fire this morning that he had come home early and gone back to sleep in his own bed. By the time she realized it, it was too late."

The young firefighter glanced at Irik, then at the ruins of the house. "When we . . . when we got to him—"

"I know," Irik snapped, not at the firefighter, but at the situation. "It was too late."

The fat woman burned her son to death in a fit of rage directed at her boyfriend. Probably the guy was out sleeping in some other woman's bedroom now. He wouldn't know yet what had happened—and when he found out, he likely wouldn't care.

How could a mother ever escape the guilt of what she had done?

Lieutenant Irik's job wasn't that of the world's conscience. People had to deal with the skeletons they hid in their own closets. The tall investigator walked slowly away, stood next to the canal and gazed into the brackish water, waiting for the house to cool, for the little boy to be carried away like an overdone roast. Irik's job was to reconstruct, identify, and preserve evidence of arson, and then to testify.

He wasn't the world's conscience, although maybe the world needed one.

Later, underneath the winter-gray sky, with sea gulls crying overhead, he stood alone among the blackened debris in the house's destroyed bedrooms.

Shovel in hand.

43

THE MISSISSIPPI RIVER FLOODED ONE YEAR, ACCORDING TO a joke told around the firehouse. A farmer confronted by the rising waters ignored warnings to evacuate.

"I'm staying in my house," he said. "God will take care of me."

The flooding waters lapped at his doorsteps. A neighbor speeding by in his pickup truck heading for high ground offered the man a ride to safety.

"No," the man declared. "I'm in the hands of God."

Water poured into his house. A boatman offered to rescue him.

"I have faith," the man said. "God will rescue me."

The waters rose. The farmer climbed to his rooftop. A helicopter pilot spotted the man clinging to his wind vane. The helicopter hovered. "Get in," the pilot encouraged. "I'll save you."

"No!" cried the stubborn farmer. "I'm waiting on God to save me."

Flood washed the house downstream. Clinging desperately to the wreckage of his house in the swirling waters, the farmer cried to Heaven: "God? God, why has thou forsaken me?"

Came a voice from above: "What else do you want? I sent you a pickup truck, a boat, and a helicopter."

Firefighters liked the moral behind the joke. They felt it applied to them in particular. God might indeed look out for firefighters, as Jim Barrett asserted, but He looked out for them by giving them the ability and the intelligence to look out for themselves in a fire.

Firefighters got into trouble when they began treating fires as routine, when they failed to use their God-given abilities to keep their senses honed to the humbling fact that man's oldest enemy could destroy him almost on a whim.

Lieutenant Rod Harris rediscovered that truth on a routine pushout to a fire in an old, ugly, yellow troll of a hotel near South Beach, on the outskirts of the built-up yuppie area of designer shops and high-rise condos. The furniture in the lobby caught the flame first, either from a dropped cigarette or the Christmas tree lights shorting out. The blaze stepped a few toes into the dirty carpet, then retreated and ate its way slowly, almost reluctantly, up the back of a worn sofa, like a young boy testing for a swim in cold water. Unseen, it gnawed into the wall and journeyed toward higher reaches. By the time it was discovered, flames licked around the white ceiling plates.

Harris, the officer on the scene, dismounted his apparatus to assess the situation. He scrutinized the flames through the picture window that gave a wide view from inside of the street and the deli on the opposite corner where a group of lunchers gathered to check on the excitement. Visibility was good. There appeared to be low heat and almost no smoke. Mike Brady joined Harris. They stepped inside the lobby together for a closer look, contemplating using hand extinguishers instead of breaking out the hoses. You didn't use a .270 Winchester to swat a gnat.

"You didn't think to bring the marshmallows, did you, Lieutenant?" Brady quipped.

Harris removed his mask. Masks and air bottles were virtually unknown luxuries in the "old days," back almost to when fire helmets were made of leather and boots of rubber that sometimes melted in contact with embers. Brady grinned.

"Lieutenant, you look like Robert Duvall in *Apocalypse Now*. Know what I mean?" He sniffed the smoke as the character had in the movie, exaggerating. "'I love the smell of napalm in the morning.'"

"Leave the comedy to Merker, Brady."

Brady laughed.

Harris eyed the blue flames licking delicately around the ceiling plates. They popped merrily and busted pretty little bubbles, like a kid chewing a jaw full of Fleer's bubblegum. They appeared so harmless, so innocuous, that Harris stood a moment, entranced by their beauty. They reminded him of an electrical fire call in a hotel ballroom.

It hadn't only been Harris who was mesmerized by the ballroom display. He and all the other firefighters, the hotel manager, some of the hotel employees, and maybe even a guest or two, stood in the great ballroom with their heads thrown back, wonder written on their faces. They could have been watching the landing of a UFO.

The ballroom had a high ceiling with great, crystal chandeliers. Among the chandeliers, reflecting in them, played nimble balls of blue and green lights, like foxfire or free-floating neon. Merry little lights, skipping and playing dodge and dancing on their tiptoes. They painted the ballroom in wonderful lively pastels.

"I've never seen anything so beautiful," the manager whispered, awed. "What is it?"

"I haven't a clue," admitted Harris, who in all his years combatting fires had never witnessed such a spectacle himself. Nothing was burning; there was just the amazing light show.

The manager shook his head to clear it, as though coming out of a hypnotic trance. "I'm getting the hell out of here," he said.

The display ended when Florida Power & Light came and doused the power. Harris had not felt threatened by the show, no more than, now, in the lobby of the old yellow troll of a hotel he felt menaced by the blue flames in the ceiling. The merry flames seemed to be laughing, playing. Tiny little harmless creatures with sunny natures.

Harris's firefighters doused the lobby fire with hand chemicals and stamped out fire patches with their big bumber boots. Harris directed the laying of a booster line from the engine to the lobby, as backup in case the fire in the ceiling proved to be more extensive than it appeared. You couldn't always judge a fire by its cover.

The ceiling *breathed*. That caught the firefighters' attention. It seemed to inhale, rattling the ceiling squares. An instant before the explosion, Harris knew he had been lured into complacency. You never trusted a fire, he knew that. God looked out for firefighters, but He also expected them to look out for themselves.

"Incoming!" someone shouted, mimicking the combat warning for an artillery shelling, as a tremendous fireball demolished the ceiling. Flaming tile shot from the center of the fireball like shrapnel. Heat curled wallpaper the length of a hallway that opened off the lobby. A tremendous flash of light temporarily blinded the astonished firefighters as they threw themselves—or were thrown—to the floor.

Harris grabbed his head and held it against the blast, half expecting it to fly across the room and slam into a wall, as his helmet had.

The explosion filled every corner of the room. Then it sucked in the picture window on implosion. Shards of deadly glass sliced through the smoke. They spanged into the walls like darts. Glass pattered around the prone firefighters.

And then everything turned quiet. One by one firefighters regained their feet and looked around in silent awe, checking out each other and their surroundings. Harris stared, amazed that no one had suffered injuries, even more amazed that the explosion had extinguished the fire in the ceiling. Astounded firemen stamped out the few fallen ceiling tiles still burning on the floor, then stared up into the exposed rafters and joists. The attic was filled with smoke, but there was no color. The fire was really out.

Later, Harris figured out that heat building up between the ceiling and the roof caused the explosion. The blast put out the fire.

"I sent you a pickup truck, a boat, and a helicopter," someone murmured. "But finally I had to do it myself anyhow."

44

FIRE RESCUE WORKED AUTO CRASHES, ACCIDENT INJURIES, choking babies, and heart attacks. Fire Rescue also worked the crime casualties—barroom fight injuries, wife batterings, shootings, stabbings, babies boiled in water, little old ladies tortured for their life savings.

"Don't it make you feel bad when you see somebody die and you can't do nothing about it?" little Manny the semi-orphan asked once, pausing with the firehouse basketball to reflect. That was before Manny's last Christmas at Station 2, before he grew up or something and took to the streets full-time.

"I seen somebody die once," Manny said. "I don't never want to see it no more."

"You can't save 'em all—but you can save some of 'em," Merker or somebody responded.

It was a phrase Jim Barrett coined several years before and the fire paramedics adopted as their informal slogan. *You can't save 'em all—but you can save some of 'em.* At that time, medical doctors still rode Fire Rescue with the medics. Barrett had a new doctor riding the van with him, a female who began her medical career in New York City emergency rooms. The emergency rooms, Dr. Nelson said, were such zoos that she soon quit to enter private practice. The boredom of private practice drove her to become a "love boat" doctor on a cruise ship docking out of Miami. That didn't last long either. She finally answered a journal ad to become a doctor with the paramedics of the Miami Beach Fire Department.

She was to discover in a most dramatic way that although she had treated combat wounds from Hell's Kitchen, Combat Alley, Fort Apache in the Bronx, there was a huge difference between handling a patient in a sterile emergency room surrounded by all the special accoutrements of medical science and treating one in a dark alley where rats peered at you from the fringes of your flashlight beam.

"Jesus God!" she exclaimed the first time she ran on a gunshot call with Jim Barrett. She crouched with him at the mouth of a darkened alley while somewhere in the shadows rats rattled garbage cans and a man lay wounded and possibly dying. He moaned softly. Chills tripped along Doc Nelson's spine.

"Jesus God," she said again, trembling.

Response time for Fire Rescue averaged about two minutes anywhere on the Beach. Response time for the overburdened police averaged six or seven minutes. That meant Fire Rescue often arrived on a scene ahead of police. Even though the paramedics carried bullet-proof vests, and some, like Ed Delfaverro, even went armed, police cautioned them to never enter a crime scene unescorted. They might lose their own lives in the process of trying to save the life of someone else. In a dark alley, a Fire Rescue uniform and badge resembled a cop's. Paramedics did not receive combat pay.

Besides, some people in the streets really didn't give a damn about the difference.

New York or not, Barrett's doctor did not comprehend the streets and their danger. She started into the dark alley to help the wounded man. Barrett grabbed her. Gunfire had erupted in the alley just minutes before.

"You don't know what the hell's down there," Barrett warned.

That shocked her back to reality. If a guy like Barrett wouldn't go into an alley, then anyone else was a fool to try it. Dr. Nelson, pale from the tension but steeling herself against the casualty's moaning, jumped behind cover and crouched there next to the paramedic.

"This is the way it's done," Barrett explained. "We wait for the police. I don't want to take a chance on losing my doctor. You're not even paid for yet."

Police arrived and, drawing arms, quickly secured the alley. Barrett and his doctor ventured forward. Their broad-beamed flashlights picked out a young man sprawled facedown next to an overflowing Dumpster. Unconscious now, with a large-caliber gunshot wound to the gut, he floated in a dark pool of his own blood.

Barrett and Nelson applied basic lifesaving procedures—check and clear airway; monitor BP and expand the arteries and veins with an IV lifeline; oxygen; pressure bandages and drainage; shock treatment.

Doc Nelson was visibly shaken. "There's so much we can't do out here," she cried.

"We stabilize and transport," Barrett said. "We're *it* out here in the streets. There ain't nobody else but us. We see it all—and we have to get down and dirty with what we have."

"He's dying."

"We can't save 'em all—but we can save some of 'em."

For his own part, conscientious and intense as he

was, Otto Ramirez accepted Jim Barrett's assessment that you couldn't save 'em all. The old rotted residue of blood in the Rescue 2 unit still emitting its odor of death reminded him of that. Christmas was only a week away. Seven more shopping days for Barrett to peruse his catalogues.

Not everyone was going to make it to Christmas.

"Otto, you have Christmas Day off this year," Ramirez's wife reminded him. "It's going to be a family day with the kids. We're staying home. We're not going to the station house or anything."

Ramirez looked forward to having Christmas off. The month had worn on him, with its Silly Season and back-to-back rescue calls. He got off shift at seven A.M. after a sleepless night running calls to drunks falling off the curbs and the like, and arrived at his nursing class by eight. As soon as classes ended, he rushed to hold down his second job as a dive technician. Study after that and a few hours sleep and classes the next day, then back to the station. No wonder his wife sometimes complained.

Mogen with his elf's hat, and Steve Hoffman bleating the airwaves with his toy doll's cry, and even Stu Merker's unfailing good humor, failed to draw the Rescue man into the general high cheer that pervaded Firehouse 2.

"Know what's wrong with you, Otto?" Daugherty decided. "You're too serious. You're a Scrooge."

Ramirez tried to lighten up. "It's beginning to look a lot like Christmas . . ." he sang to himself, cheered by the thought of having a holiday off.

His eyes lingered on a pine tree decked with lights in the yard of a residence the van passed on a call-out to check on the well-being of neighbors. The colored lights momentarily cast back the gloom of approaching darkness and washed pastels across the faces of

Merker and Daugherty, who crewed Rescue 2 on Ramirez's shift. Merker had been wheedling Captain Garcia to return him to working fire suppression. Fire Rescue could be depressing, especially at Christmas.

"Betcha it's a Code 45, a DOA," Daugherty decided.

"Mr. Gloom and Doom," Merker shot back.

Police were parked in front of a pleasant-looking little Florida bungalow with a cedar wreath on the front door and more cedar-and-holly wreaths in the windows. The house appeared dark inside, except for the dim glow of colored Christmas lights.

Officers knocking on the door received no response. They shouted for Ted and Ellie. The house slapped the names back.

"We live next door," a bald man explained to the police officers. "Last night my wife and I were in our bedroom, which is on this end of the house. We heard something that sounded like a scream. My wife heard what she thought was somebody falling down or knocking something over. I don't know. We've been knocking on the door, but Ted and Ellie don't answer. We've even tried to phone them."

The door was locked.

"I'm sure they're in there," the bald neighbor insisted. "Their car is in the garage. They're getting old. They might be ill or something."

"We have to go in," a policeman decided. He reared back. The wooden door splintered beneath the big foot he thrust through it.

He entered the house quickly, calling out the names of the occupants. He returned momentarily, his face grim. He motioned to the waiting paramedics.

"Check them out," he said, "but I think it's a waste of time. They're dead. Looks like someone broke into the house through the kitchen door."

Ramirez dreaded the thought of viewing more corpses. He didn't feel like singing anymore. Even Merker seemed subdued as the medics followed the cop, trudging through the oppressed stillness across a living room lighted by a Christmas tree, down a short hallway, and into a kitchen bathed in fluorescent light. Two bodies lying facedown on the tile floor spoiled the kitchen's neatness, as obscene as feces on a plate.

Tim Daugherty cast a sharp look at Ramirez and let his breath out in a sharp wheeze. Merker froze. He looked at the old couple lying on the floor. They looked like grandparents. He looked away. Ramirez steeled himself and dropped to his knees. He checked pulses. There were none.

"Dead—for several hours," he guessed.

The flesh around the old couple's necks was swollen, bruised, caked. Both man and wife had been manually strangled to death and their hands tied behind their backs with cords ripped from venetian blinds. A trickle of dried blood, bright and dark at the same time against the pale tile, drained from the woman's nose.

What appalled and stunned cops and paramedics alike, angered them, what was more obscene than the deaths, was the manner of the deaths. Ted and Ellie wore matching slacks. Their slacks were jerked down to their knees, exposing their pasty white buttocks to the fluorescent light and the professional scrutiny of strangers. Obviously, both old people had been raped anally, a fact corroborated by autopsy.

Nothing else in the house appeared missing or disturbed. Perpetrators, "perps" in the lingo of cops, some evil and perverted creatures of the night, had clawed their way into these old people's castle to satisfy some demented craving. Ramirez found it almost impossible to imagine anyone committing such a monstrous act.

He rose silently to his feet, helpless in his growing outrage. No one said anything. Daugherty and Merker followed him out of the house. They paused at the Christmas tree. Just below the star at the top dangled a plastic nativity scene of Joseph and Mary and the Baby Jesus. Considering what had occurred in this house, the simple message it bore was ripe with bitter irony: GOOD WILL TO ALL.

In spite of what Ramirez had witnessed with Fire Rescue—the violence and stupidity and carnage, man's inhumanity to man, the casualties of human endeavor and progress and failures—he was not by nature a cynical man. He let Merker drive the van back to the station. Each remained cloaked in somber thought. They found little cheer in the bright lights of Christmas. Ramirez experienced a great sadness welling inside, a remorse that at this most holy time men still managed to do really horrible things to each other.

Maybe, he thought, there was no question of saving 'em all. Maybe there was no salvation for any of 'em.

45

EVEN THE HOG'S HEAD CHIEF WITH HIS SANTA'S BEARD seemed to gain weight over the holidays, simply from the aroma of all the goodies.

"I'm getting so *fat*," blond Glenda Guise lamented, which she wasn't.

Mogen suggested she take vitamins to supplement her uncertain diet during Christmas. Merker suggested Mogen might lose twenty pounds of ugly fat—"or maybe just an ounce or two"—by completing the circumcision he was always talking about. Barrett announced he had discovered the perfect gift for his wife, then changed his mind immediately afterward and said he hadn't after all. Vance Irik complained that he still hadn't captured Miami Beach's elusive hotel arsonist. No one had been killed or even injured so far in the torched hotels, but Irik felt it to be just a matter of time before one of the flames turned Stage Four.

Christmas always proved a nostalgic time at the fire station. Losing one of the firefighters to retirement or death was the same as losing family. The firefighters remembered that one family member had to retire because the blowing sirens made him deaf. Lenny

Rubin choked himself unconscious on smoke in a basement during a hurricane fire and then drowned, partly because the early air-breathing apparatus sometimes malfunctioned. Even some of the kids and neighbors who used to hang around the station were missed and remembered.

"You know," a firefighter mused, "that little kid Manny would be around here at Christmas all the time."

"It's not the same anymore since Jim Reilly died," Captain Luis Garcia said in his precise politician's tone.

Reilly was Station 2 legend. During his long career, before he died suddenly of cancer, he worked trucks and engines and helped pioneer Fire Rescue and the fire prevention–arson investigation office where Irik now worked. Reilly had fought many of the legendary fires discussed and replayed and used as training examples for probies in fire school. Big deadly blazes like the ones at Manny's Restaurant and the Concorde Cafeteria.

There was another darker side, however, to the gentle nostalgia that infected the firehouses at this time of year. Firefighters lived with their harsher memories as well, having witnessed death and tragedy as only they and cops and soldiers in combat witnessed it. Sometimes Gene Spear, now retired, awoke at night bathed in a cold sweat from nightmarish images flashing through his sleep. On bad nights his sleep was made restless by ghosts of suicide victims with the backs of their skulls blown off and their brains scrambled all over the floor and walls; fire-burned corpses—crispy critters—with their arms and legs charred down to nubs, like the pointed ends of burnt logs; children thrown through windshields;

floaters in Biscayne Bay, bloated and gnawed on by crabs, with the skin slewing off; murder victims sliced and diced and machine-gunned. . . .

And there were the fires he had fought too, where the flames and heat were so intense they scorched the lungs with one breath if you dared remove your SCBA. Other people in their nightmares were chased by monsters or snakes. Sleep trapped Spear inside his twenty-five years of fire-fighting. Blazes, burning infernos, threatened to annihilate him. Fiery roofs dropped on him. Smoke filled his lungs, robbing him of breath, poisoning him, blistering his exposed skin.

"Honey, honey, wake up," his wife sometimes cried, awakened herself by his nocturnal suffering. "You were gasping for breath. Can't you breathe?"

In the years to come, Jim Barrett, Glenda Guise, Otto Ramirez, and the others would all have their nightmares. And some of those nightmares would come from the stories they had heard told by Gene Spear and Jim Reilly about the Concorde Cafeteria fire on Collins Avenue and Nineteenth Street. The fire occurred before most of the younger firefighters had joined the department, but it was a fire that had become legend and was often discussed at fire school and used as a training aid.

Gene Spear described the fire so vividly that the younger firefighters felt as though they had actually been there. Some of them had tears in their eyes. They envisioned the old Concorde as it must have been then, a restaurant frequented by the Beach's large and predominantly older Jewish community. When fire broke out that fateful evening, the restaurant was crowded. Haunting how many of the diners were elderly Jews who in their youth had eluded Hitler's final solution, only to end up, these many years later, massed writhing into a single tiny bathroom where heat and lethal gases caused by smoke threatened to do to them what Hitler had failed to do.

46

GENE SPEAR REMEMBERED THAT AN ARSONIST STARTED THE Concorde Cafeteria fire and that he went to prison for murder. He couldn't remember the arsonist's name now nor why he started the fire. These points of information had faded into the general memory of so many other pyros who, for varying insane reasons or, just as insanely, for no reason at all, had committed the crime of arson.

Spear fought the fire, and later learned that some disgruntled patron, one of the walking psychopaths cities produce, a former employee with a grudge, *someone*, had darted into the cafeteria at its busiest time armed with a gallon jug of gasoline. On most afternoons and evenings around dinnertime the cafeteria crowded in more than one hundred people. The guy ran gleefully among the stunned diners dousing the place with gasoline, like some mad gardener sprinkling his cabbages with water. Then he laughed loudly just before he struck a match, held it aloft so that it reflected in his eyes and against his teeth, and flung it at the spilled gasoline. The restaurant flashed immediately into one of Hell's many pits. The arsonist was one of very few to escape completely unscathed.

The pushout reached Firehouse 2 on a three-alarmer. This was no kitchen grease fire. Even as the pumpers and ladder trucks from three firehouses, supported by Fire Rescue vans and a squad of police, converged on the cafeteria with sirens blowing, busting traffic lights, black and gray smoke pumped hundreds of feet into the late afternoon sky. It could be seen from as far north as Fort Lauderdale, as far south as Homestead.

From down the block as the front area of the restaurant became a siren and red light circus, Spear glimpsed flames leaping twenty feet into the air, cracking and snapping like rabid things. The entire building had become almost immediately involved.

The first thing Spear focused on as he unassed the apparatus were the large plate-glass windows that faced busy Collins Avenue, noting in them the dervish play of shadows in smoke created by the dancing flames. Before his eyes, before he could react, an elderly man crazed with terror crashed out of the cafeteria through one of the windows, trailing smoke and shattered glass and a burst of heat as from the suddenly opened door of a giant oven.

Dazed and bleeding, as crazed as an animal, the old man crawled away from the cafeteria on his hands and knees. Spear and Luis Garcia reached him. He fought them off as though, in his blind fear, they were the fire taken human form. The firefighters grabbed him underneath his arms and half dragged, half carried him to safety.

It was for fires like this that firefighters drilled, so

they could act automatically and efficiently, keep their heads while all about them others were losing theirs. Pumper crews quickly laid attack pipe to force a corridor into the boiling cafeteria, while truckies organized quick penetration and rescue teams, as well as smashing the remaining plate-glass windows and chopping holes in the roof to bleed off the building's heat charge. Pressure would blow its own exits anyhow, creating flashovers and backdrafts.

On the periphery of his vision and hearing as he worked, dogging a hose fitting to a hydrant, Spear saw and heard police and Rescue men running about, gathering apparent survivors from the fire and corralling them off to one side for treatment. There was a lot of yelling, screaming, and crying out. Frantic people pointed at the blazing cafeteria.

"There are others in there!" they shouted. "Oh, my God. *My God, they're being burned alive!*"

Someone else screeched: "He ran in and threw gasoline on us!"

Firefighters were too occupied at the moment to worry about how the fire started. Catching the pyro was the job of the police and arson investigators. Firefighters concentrated, first, on rescuing victims, and then upon containing the fire and killing it.

That was job enough.

Fire officers decided upon an interior first-in assault to reach survivors, backed up by external hoses. The outside hoses were already shooting water through windows and doors, trying to cool down the structure. Steam belched skyward with the smoke. Spear laid a 3½-incher with Rod Harris and maybe Luis Garcia. He had just opened the nozzle, probing with a hard bar of water toward the front of the cafeteria and the smoke boiling from the smashed-out windows, when word came that his crew had been ordered to first-in.

"There are still a lot of people in there," Lieutenant Jim Reilly told him.

Rescue squads at the door, prevented from entering by the intensity of the flames, grabbed survivors as they stampeded into the fresh air. Some of them had their clothing burned off; others were still smoldering, their charred and smoke-grimed faces frozen in death masks of fright and pain, weeping from sheer panic and the acidity of the smoke. Firefighters, coated and helmeted and masked in stark contrast to the near-naked victims, half dragged the survivors away from the conflagration and into the fresh biting air, where the elderly men and women wheezed at it, sucking like bellows that could not fill their lungs because of holes in them.

Spear on the nozzle and the two firemen on the hose with him worked their way into the restaurant. Distance and layout were difficult to judge in dense smoke. Tapping over surfaces with one gloved hand, wrestling the water-forced hose with the other, Spear felt his way inside like a diver submerged in the muddy Mississippi. His stream of water bored its way through the smoke. He feared he might strike other survivors with it—but there would be no more survivors unless he cooled off a rescue path through the flames. Besides, by now anyone still alive and able to negotiate on his own had already bolted to safety. Human nature had taken all the others into hiding. Either that or they were dead or had succumbed to the heat and smoke.

The powerful water jet slammed dishes and furniture about in the smoke. China shattered and forks and spoons ricocheted off walls. While Spear's crew concentrated on beating back the flames, truckie rescuers plundered for survivors. Spear became peripherally aware of other activity around him in the dense smoke as firefighters made survivor grabs and ushered smoke-inhalation and fireburn victims out of the flames. He heard crying and coughing and shouts.

"Glory, glory to God!" someone cheered.

"There's supposed to be more people hiding in the kitchen and hallways," a truckie shouted at the enginemen on the hose. "We need to work over there and toward the back."

Spear sensed a hallway, felt his way along it until he saw flame color ahead in what he recognized as the kitchen. He glimpsed a freezer next to a row of stove burners. A burning table directly in front of him collapsed into a bonfire. A thin finger of flame reached from the rubble toward the ceiling, joining fire dripping down from above.

Nozzle turned to spray instead of to stream, Spear drenched the table into submission, then tipped his weapon toward the crackling ceiling. His muscles ached from the hose recoil, scalding sweat dripped into his crotch and soaked the uniform he wore underneath his turnouts, but he fought on against the super heat and the licking flames that surrounded them until he tasted the bad air at the bottom of his bottle.

Forced outside to replace his air bottle, he discovered a scene of utter pandemonium whose excitement and tension reminded him of TV documentary scenes of WWII firestorms in London—smoke; rubble in the street; uniformed men rushing about and shouting orders; casualties lying moaning on blankets; Fire Rescue vans filled with victims coming and going with sirens blowing and red lights flashing.

A man with tufts of what had once been white hair, but were now black and oily from soot, rose feebly from a blanket and in a heartrending gesture reached a pleading hand toward the fire.

"My . . . my wife . . ."

Spear charged back into the fight with a fresh air bottle. At the smoke-shrouded cafeteria door he bumped into a firefighter carrying an unconscious woman across his shoulders. The firefighter stumbled but maintained his balance. Spear reached to help.

"No . . ." the masked firefighter gasped. "More . . . there're more inside."

As Spear followed his hose into the acid smoke, he heard someone call out. It came from his right, deeper into the conflagration. A firefighter. "Oh, Holy Mother of Jesus! Need some help over here."

Spear automatically responded, sloshing blindly through floor water and soot, feeling his way toward the desperate voice, past the inferno's reaching flames and through them to a swinging metal door leading out of the main dining area into a hallway. Smoke disoriented him.

"Keep talking," he yelled to the unseen smoke eater. "Where are you?"

"Here. *Here!* There's a toilet."

The voice was so near he could have reached out and touched its owner. The smoke thinned slightly, enabling Spear to see the truckie and several other firefighters clustered around an open door, staring in alarm. A sign on the door said BATHROOM.

Spear looked into the bathroom, so small it contained only a single commode, a urinal, and a sink, and he understood. Before his startled eyes played a scene of such Dantelike horror that it immediately seared itself into his brain, to be replayed forever afterward in his nightmares. About fifteen elderly people, some of them Holocaust survivors wearing Hitler's tattoos on their frail arms, had in a stampede of panic jammed themselves into the windowless lavatory. A mass of writhing, twisting, clawing human flesh filled the closet-sized room. Choking on smoke, passing out from heat, the old people had lost their senses and in panic and desperation clawed at each other and climbed onto each other, the stronger stomping on the weaker. There was hardly room for the unconscious and the already dead to sink to the floor. Arms like the weaving heads of snakes in a nest of snakes reached imploringly out of the dying clump of people as it collapsed and spilled out the door at the feet of the appalled firefighters.

An old woman wearing a patterned blue dress crawled across the floor and grabbed Spear's boots. Spear said—and it would have sounded corny under any other circumstances, but he meant it—"We'll save you."

As on cue, firefighters plunged into the mass and each hoisted a body onto his strong shoulders. Spear lifted the old woman wearing the blue dress into his arms as easily as if she were an infant. Fortunately, most of the people were elderly and as frail as skeletons.

"Stay inside the bathroom!" Spear shouted, promising, "We'll be back for the rest of you."

Those left behind were too far gone to respond.

On his way out, Spear ordered a crew entering with a small red line to defend the bathroom until rescuers returned with reinforcements.

Fighting exhaustion, the rescuing firefighters of the first wave stumbled into the fresh air with their survivors, then led additional truckies back into the maelstrom to extract the remainder of the Concorde Cafeteria patrons. When they returned again to safe territory, each emerging from the smoke with an unconscious or, at most, semiconscious victim on his back or shoulders, the long file of them resembled knights or soldiers braving Hell to bring back the newly damned.

Search teams reported that with the rescue from the lavatory, the cafeteria appeared clear of bodies, both live ones and dead ones. The fire was at Stage Four, burning freely and fiercely and resisting all efforts to halt its advance. The cafeteria was lost. The fire captain reluctantly ordered a retreat. Just as reluctantly, troops pulled back to a defensive perimeter. The order came down to surround and drown, let it burn but cover exposures so that it did not spread to adjacent structures.

Five people perished in the Concorde Cafeteria arson. At nights, sometimes, retired firefighter Gene Spear saw their faces among the elderly trapped inside the bathroom as they writhed and screamed for salvation. And, through him, even the newest probie saw the same faces.

47

FIREHOUSE 2 PREPARED FOR ITS TRADITIONAL CHRISTMAS dinner. Wives and girlfriends breezed in and out and stood on chairs to hang wreaths and paper bells and cutout Santas. Laughing, they bickered over the menu, agreeing at Jim Barrett's insistence only that the main course not be Steve Hoffman's meat loaf nuked in the station's new microwave. The station filled with such teasing and laughter and cheer that it appeared to be about to burst at the seams. Chapman jerked Mogen's elf hat down over his ears. Merker stood on a chair to announce that Barrett had at long last decided upon a gift for his wife.

"No! No, he hasn't. False alarm. Tune in tomorrow for an update of this continuing saga."

Daugherty produced a pair of rusty hedge clippers for Mogen's circumcision and chased him around the garage with them. Merker and Harris conspired to steal Hoffman's doll crier. Even Vance Irik, worn and discouraged over his hotel arson investigations, cracked a grin or two, and he and Captain Garcia were spotted laughing together over some joke.

In the midst of all this hilarity, the crews of the two Rescue vans, Rescue 2 and Rescue 2-2, were kept so busy they seldom had time to join in the festivities. On a particular day the week before Christmas, Otto Ramirez and his crew assumed first-out with the beginning of shift at seven A.M. Glenda Guise's Rescue 2-2 would take over first-out at seven P.M. for the remaining twelve hours of the shift.

Not that first-out or second-out meant a bunch during the Silly Season, when *both* units frequently took simultaneous calls.

"Rescue Two, Rescue Two, ill person at . . ."

Merker and Daugherty hit the stairs with Ramirez. Ramirez had had time for a quick workout in the station's weight room. It was too much to ask that he be able to crash for an hour or two and watch TV in the ready room beneath the glinting glass eyes of the Hog's Head Chief wearing its ridiculous fire hat and Santa's beard.

Neal Chapman, playing cards at a table with fire

suppression, called out as Rescue 2 darted for the van and the station's big doors rolled up: "Show time. Otto's got to make a living. He owes the coffee fund."

Hoffman, on Guise's crew which was still at the firehouse, bleated his doll's cry over the radio, heckling, as Ramirez whipped the van onto Pine Tree and followed it to Forty-first Street and then east to Collins, siren pulsing. His amusement, however, was short-lived, as his crew then received a call: *"Rescue Two-Two, Rescue Two-Two, man at doctor's office complaining of chest pains...."*

Guise stared at the intercom. *"What?"*

Rescue was being summoned to a *doctor's office*?

Guise, Hoffman, and John Creel raced to their van. Merker in Rescue 2 blasted them with a doll's cry over the air, heckling back.

In the Rescue ready room, sitting at his desk, Captain Luis Garcia rolled back his eyes and shook his head. One after the other, in a medley of human tragedy and folly, accompanied by the bleating from toy dolls, Firehouse 2's Rescue squads ran their endless missions of mercy.

48

A DIMINUTIVE GRAY MAN WITH BROWN, UNPRESSED TROUsers pulled up to his armpits opened the door to Ramirez's Rescue 2 crew on the fifth floor of an oceanside high rise. Behind him in the living room, perched on her chair like a sick sparrow, his wife stared uncomprehendingly at the paramedics.

"Since last night she is this way and she don't move," the old man said in English heavily accented by somewhere in Eastern Europe. "Since a long time she don't walk. Now she won't get out of her chair."

The sparrow drooled, made sparrow sounds. Her tiny hands resembled bird claws. The little man introduced himself as Albert Stein.

"Twelve years ago it was I make the choice to stay with Dottie when the end comes," Mr. Stein continued. "She is not so easy to take care of, but I am take care of her. At eighty-one years old I am, I am staying with her every day since she was sick. It is a dreadful disease, the doctor says. Alzheimer's disease, he tells me. 'Mr. Stein, I am so sorry,' he says, 'but as the disease she grow more severe, your Dottie may not even recognize you her own beloved husband.'"

More weak gibberish from the sparrow.

"She don't make so much sense," her husband

admitted, then added with pride that, Alzheimer's or not, she still remembered her devoted mate. "He was wrong, the doctor. My voice Dottie still knows."

He held his wife's hand tenderly, saying he called Rescue because he was worried about his wife's heart. Stu Merker pressed the bird's claw, spoke to her softly. Her vacant eyes told him nothing was filtering through to her crippled brain, that nothing might ever filter through again. After a point, Merker recalled from other experiences, Alzheimer's patients did not experience good or bad days, simply days with varying degrees of fog. After a while they lived in a world that existed solely inside their own heads. Maybe they didn't even exist there.

"Some days she looks like she can hear me, other days not so good," Mr. Stein apologized.

The paramedics decided to transport her to the Miami Heart Institute, where she was an outpatient. Maybe the doctors there would know what to do with her. Daugherty and Ramirez stretchered her out to the van while Merker helped Mr. Stein totter alongside, where he gripped his wife's hand as though afraid of losing her. She appeared so frail, her heart so weak and her blood pressure so low, that the paramedics set up an IV in the van for her.

Mrs. Stein emitted a piercing scream when she felt the needle prick her arm, her first reaction to any outside stimuli that the paramedics had observed. She flailed weakly at the O_2 mask, trying to rip it from her face. The surprised paramedics attempted to restrain the tiny creature without harming her brittle bones or bruising the thin skin that covered them.

"Maybe with her I can be of help," Mr. Stein offered, looking worried. "Maybe with her I ride? She is so very frightened."

Ramirez helped the little man into the van. Mr.

Stein gently took his wife's hand. His loving gaze never left her shriveled and vacant face. Although she made no other indication of his presence, she immediately calmed and babbled softly. She resisted no further as paramedics taped in the IV needle, administered oxygen.

"You see? You see?" Mr. Stein chortled. "My Dottie she knows me when I am here."

He smiled, pleased.

In the meantime, Guise's Rescue 2-2 rolled on its "chest pains" call to a medical center next door to a Miami Beach Hospital emergency room. Not only that, but the call was to a cardiologist's office. Glenda Guise and her crew rushed a collapsible stretcher upstairs in the elevator. Hoffman complained mildly, shaking his head and shrugging.

"I guess I ought to expect chest pains in a cardiologist's office," he said.

"Where else?" Creel asked.

The cardiologist himself rushed from his examining room. He seemed so grateful for the paramedics' arrival that he could have hugged them.

"A patient came in complaining of chest pains," he explained. "He's still irregular now."

"He had chest pains and you told him to come to your office?" Guise demanded.

Normally, chest pains rated 911 and a fast trip to an emergency room. Hoffman the cherub, who looked as though he escaped from "The Waltons' Christmas Special," would have offered his usual acerbic comment, except Guise cut him off.

"Have you started an IV?" she asked the doctor as he hustled them to the examining room. A 3F man—fat, flatulent, and over forty—lay bare-chested and beached on a table.

"Rescue Two-Two, Rescue Two-Two . . ."

Hoffman rolled his eyes.

". . . Possible cardiac arrest. Emergency room entrance . . ."

"What?" Hoffman cried. They were standing across the street from it, catching a fast breather.

Guise shot glances at her partners. They grabbed their bags and ran across the street to where a man lay collapsed on the steps of the emergency room entrance, no more than fifteen feet from the doorway. The guy appeared in his fifties, with a face as pale as Dracula's. Sweat streamed cold from every pore. Several people surrounded him, including an E.R. nurse who knelt at his side.

"I was afraid to move him," the nurse said. "I don't think our liability insurance covers him until he gets inside."

Hoffman bit his tongue to keep prudently silent.

"We're going to get you to the *real* doctors inside," Guise assured the victim, trying to ease the bite of her own sarcasm. "No reason for us to do anything when the real thing is so close. Right, sir?"

Silent gratitude replaced some of the pain in the man's eyes.

"Unbelievable!" Hoffman chafed, letting it all out. "Doctors are so goddamned specialized and liability-conscious that they can't do a thing for a dying patient unless he walks in and lays down for them—or unless *we* carry him in."

As Guise and her crew carried their heart attack victim into the hospital, Ramirez and his crew extracted people from a gray Nissan and a black Honda that had mated with each other at the intersection of Collins and Forty-first Street. The young driver of the Nissan suffered a possible broken wrist, while her mother had difficulty breathing and complained of back pain. Rescue 2 transported both to Mount Sinai Hospital.

The paramedics enjoyed quick cups of coffee in the emergency room while they completed their paperwork. The young girl burst out of the admissions office cradling her broken wrist in her other arm. She was crying and her eyes were red and puffed.

"Do you know how much money they charge just to take a look at my wrist?" she demanded, blinking at Ramirez as though stunned.

Ramirez forced a smile. "Treatment doesn't come cheap," he said lamely.

"I guess I'll just have to suffer." She swung on down the hallway to check on her mother, holding her wrist.

Later, Glenda Guise's crew ran on a stabbing between two homosexuals. She knew she would be teased over it.

"A Christmas stabbing," she wisecracked. "Isn't that a tradition for everyone? Oh, God. I'm getting as cynical as the rest of this firehouse. They'll have to commit me."

The stabbing occurred at a cheap rooming house, a dump of a room upstairs, laid in stained carpet that had once been orange and enclosed in walls painted a putrefying lime-green. The place reeked of cleaning chemicals, mildew, urine, and age.

"As if the color alone wasn't enough to make you nauseous," Guise noted.

The fire department had a policy that a lieutenant must back paramedics on every call involving violence. Lieutenant Ed Delfaverro showed up at the rooming house where a lithe young man with the stereotypical lisp and fluttering wrist directed the paramedics and a pair of uniformed policemen to his bedroom. There on the bed lay a second man, facedown, covered partially by a sheet. He turned his soft face toward the intruders and fluttered a hand. Everyone stood around the bed and looked.

The guy still on his feet flapped his arms in exasperation. "We had a little tiff," he lisped. "I didn't mean to hurt him, I really didn't. He knows I have a temper, the bitch. It just happened."

The guy on the bed smiled wanly.

"And . . . ?" Guise prompted.

"The whore stabbed me," the homosexual on the bed inserted. "With a pair of scissors. I asked him to leave and he wouldn't. Look."

He threw off his sheet, unselfconsciously exposing a skinny bottom with a small puncture wound in one cheek. Always ready for a laugh, Hoffman turned away to hide his face.

"You should have seen Glenda's eyes," he later recounted at the station for the amusement of the other firefighters. "John, did it look to you like she enjoyed administering medical aid to that little dude's ass?"

"It's a good thing it wasn't a snakebite," Creel said.

Seeing so much human tragedy, the "irony of human existence," as Otto Ramirez once dubbed it, gave paramedics a macabre sense of humor shared by cops and combat soldiers. Black humor somehow kept them sane. It was a way of coping.

The irony took a special twist when a Toyota at high speed swept off the Julia Tuttle Causeway at Alton Road. It jumped the curb, destroyed an exhibit of plants and planters left out overnight in front of a greenhouse, and demolished four parked cars. When the Toyota came to rest, it resembled a beer can crushed in some big redneck's fist.

Pieces of human flesh as well as teeth, hair, and bone fragments smeared the Toyota's compacted interior. Blood from the male driver and his three passengers, two men and an enormous woman in the backseat, drenched the interior like someone had splashed the seats and occupants with a bucketful of it.

What was left of the driver occupied the four inches of space that remained between the steering wheel and the back of his bucket seat. He sat folded into it with his neck twisted at an angle, like the possessed Linda Blair in *The Exorcist,* who could ratchet her head around on her shoulders.

His passenger in the front seat next to him had merged with metal on impact. The dead man had knees where his ears once were, while a part of his skull rested on the dash and the rest of his head went into the glove compartment. He wasn't merely dead; he was deconstructed.

Ramirez, Daugherty, and Merker stared.

"There's not enough of him left to transport," Daugherty said.

The man passenger in the backseat apparently suffered only minor injuries. The woman next to him moaned softly. "My legs. My legs." She appeared larger than four-hundred-pound Mr. Jones who came to the Fontainebleau each December. She looked as though she could have tucked what was left of the Toyota underneath her arm and walked off with it.

"She was," Merker understated later, "huger than huge."

The paramedics determined that no one in the car had worn seat belts. The impact of the crash hurled the woman's tremendous weight around in the car, crushing the backs of the front seats and driving both front seat occupants into the dash, killing them instantly. It was like throwing a beef through the car at one hundred miles per hour.

The heavy woman became the weapon that killed two of her friends, and ultimately herself as well.

As Ramirez and his crew worked to extricate the mangle of bodies, separating the living from the dead, he stepped back a moment to catch his breath. Affected by the gore and guts, exhausted and groggy from the Silly Season pace, he caught his breath and helped load the big woman and her backseat companion into the back of Rescue 2-2 for a red light transfer to the hospital. Then, for the first time, he had a moment to look around. His shifting gaze fell upon one of the parked cars involved in the accident, a van.

He blinked, confused.

What appeared to be blood was spackled all over the inside of the van's windows.

"My God!" he screamed suddenly. "I thought they were all unoccupied."

He charged the van and flung open the side doors. His eyes darted wildly, searching for still another victim. Merker, Daugherty, someone else maybe, grabbed the paramedic, gripped his shoulders, saying, "Otto, Otto, it's okay, man. It's not what it seems."

"It's blood!" Ramirez exclaimed. "Can't you see it? We've got someone else injured."

"Otto! It's red paint, Otto. The guy must be a house painter or something. It's red paint, Otto."

Ramirez shook his head to clear it. The film that narrowed his focus gradually lifted.

"Red paint?" he asked numbly.

"Red paint," he concluded later, telling the story on himself and about how the weight of the fat woman killed her two friends.

He grinned feebly and looked around at the yule trappings brightening the interior of Firehouse 2. The theme from "The Waltons' Christmas Special" was just coming on TV. Seven P.M., time for Rescue 2-2 to go on first-out for the next twelve hours. Ramirez sighed and leaned back in his chair. He looked forward to having Christmas Day off.

"Rescue Two-Two, Rescue Two-Two, home accident at . . ."

Guise and her partners hit the stairs on their way down to the van. Merker bleated a doll's cry at their disappearing backs.

A few minutes later, just as Rescue 2 settled into "The Waltons," taking a well-deserved respite from first-out, the dispatch's voice again rang on the intercom: *"Rescue Two, Rescue Two, child drank poison at . . ."*

Otto Ramirez sighed, then bolted to his feet. Hoffman in 2-2 bleated the airways with his doll's cry. In the station ready room Captain Luis Garcia tapped a pencil on his desk. He reached for his desk mike. Then he sat back, leaving it untouched.

What the hell. Let them blow off a little steam. They were doing their jobs. They were doing them well. That was what counted.

Let the dolls cry. Let 'em cry.

49

SOMETHING ABOUT A FIRE ATTRACTED SPECTATORS; IT printed in their wide eyes as they watched in a kind of primitive wonder. Sometimes Jim Barrett thought onlookers at a fire resembled ancient tribesmen gathered around to roast meat for the first time. Sometimes, as now with the house smoldering underneath the bright yellow exterminator's tent, the superheated air inside blowing it up around the two-story house like a giant balloon, he lost patience and yelled at the spectators to "get the hell back. Don't you know it can kill you?"

Because of the tent, the fire was not sufficiently vented. It smoldered, charging itself with energy, like a man trying to control a bad temper while seriously provoked. Yellow-gray smoke escaped through several rents in the canvas, hissing as if from overstoked chimneys as it clawed its way high into the blue Florida-winter sky.

Barrett was the officer-in-charge. He stood back a moment, eyeing the fire speculatively while engine crews from Station 2 and Station 1 laid pipe and turned thick jets of water onto the house exposures, trying to cool them. The tent served as a raincoat. Water splashed off and soaked the yard.

Barrett realized that unless his troops launched an immediate interior assault, the house and everything in it was lost, gone up in smoke. Yet, he also realized that while his firefighters' SCBAs might protect their breathing poisonous fumes, they were virtually useless against other exterminating poisons that attacked through the skin instead of through the lungs.

Neighbors had no idea where the homeowners were. "They're staying somewhere in Key Largo while their house is being bugproofed," Barrett was told.

He asked the fire dispatcher to contact the exterminating company and check on the type of poison being used.

"Stand by," dispatch responded.

Firefighters couldn't stand by much longer if they hoped to save the house. Barrett paced, waiting. Chapman, who helped man one of the lines, cast him a questioning look. Interior assaults were MBFD's specialty. Chapman's look asked if they were going in—going inside that tent filled with heat, flame, thick smoke, and potentially lethal gases—or if they were going to cower outside the field of battle?

It wasn't only the exterminator's poisons. The tent itself might flash into general flame at any instant, although that immediate likelihood diminished because of the water being poured onto it. Poor venting might also cause flashovers and backdrafts.

Add it all up. Poisons, dangerous gases, backdrafts . . . It was taking a chance, rolling the dice, going into that tent of horrors and potential death.

But that was what firefighters did for a living—took chances.

Barrett checked with the dispatcher. She hadn't yet been able to contact the exterminator. He turned from the radio and reached for his own SCBA. He couldn't ask his men to do something he wouldn't do himself. His deep voice rang out above the crackling of flames.

"Get ready. We can't wait. We're going in. I'm leading it."

He cast assignments. Troops from Engine 1 would remain on exposures; Engine 2 firefighters would make the hazardous first probe.

The lieutenant from Station 1 grabbed Barrett's arm. "There are no guarantees that shit ain't as deadly as original sin," he warned. "There are no guarantees."

Barrett flashed a wry grin through his mask visor. "I never asked for any," he replied.

Now that the decision was made, he was eager to close with the enemy. The risk, he reasoned, wasn't nearly as great as it first appeared. Few exterminators used the type of poison that affected human skin.

He wondered if he might not simply be trying to make himself feel better.

Firefighters axed through the front door. Noxious smoke poured out the opening; it contained whatever toxins the exterminator used. Posed at the door prepared to lead his two lines and crews into action, Barrett felt his skin burning underneath his turnouts. Was it from the heat and the normal acid in smoke—or was it from the poison? Chapman on one of the assault lines cast him another question.

The structure was charged like a pressure cooker. The fire came roaring at the firefighters from the rear of the house, from the kitchen area. It sounded like a volcano erupting. Barrett heard glass shattering as the house popped its windows like bottles of hot champagne popping their corks.

Visibility dropped to less than two feet as Barrett grabbed the nozzle of the hose manned by Chapman and Hoffman and led the double-crew assault behind twin streams of water. Shooting at threatening flames like a door gunner on a chopper, Barrett and his patrol fought their way across the living room toward the seat of the fire in the dining room and kitchen. Merker's crew worked security and stayed slightly behind to ward off any rear counterattack. In the heat of battle they all forgot about whatever poisonous gases may have been used; they ignored the unsettling fact that the giant exterminator's tent enclosed potentially lethal smoke with them inside the house, and caused it to pump thick and heavy and hot from room to room.

Flames from the kitchen had escaped into the dining room and chewed a large hole through the ceiling, where they spread to the second floor. Even as Barrett tipped his nozzle and blasted water at the ceiling, and had it steamed and blasted back at him, he heard other flames rummaging through the second floor. As soon as he could, he turned the downstairs battlefield over to Merker and siphoned off fresh point troops from a third line to attack upstairs.

At the top of the stairs was some kind of storage room filled with cardboard crates. Firefighters axed them apart to get to the seeping fire.

Next door was a bedroom so charged with pressure that when John Creel kicked open the door, a blast of fire, like the blast from a jet engine, caught him and hurled him tumbling down the hallway. Shaken but apparently unscathed, he jumped up and threw himself back into the battle by grabbing the nearest water hose.

"It's a hot time in the old bedroom," Hoffman declared. It helped break tension.

For the next half hour assault troops poured so

much water onto the flames that water trickled through seams and cracks in the floor and sounded like a brook tunneling down the stairway. The fire retreated and died slowly with gurgling death throes. Only after it was under control and the first-in crews were running out of air in their tanks did Barrett lead the way off the battleground and turn it over to salvage and clean-up. His quick decision to work an inside assault had saved the house's roof and external frame. It could be rebuilt.

On the front lawn, crews stripped off their masks and sucked fresh air. They wiped bitter smoke tears from their eyes and drained bottles of Gatorade. Exhausted as he was, Barrett hurried off to find out about the exterminator's poison. First-in troops waited anxiously for his return.

A lieutenant from Station 1 grinned the news.

"I know," Barrett said. "You owe me money and I'm not going to be around to collect."

"There's good news and there's bad news," the lieutenant said. Hoffman wandered over, carrying a bottle of Gatorade.

"Bad news?" he overheard.

"The exterminator uses Vicane gas," the lieutenant explained. "The Hazardous Chemical Agency called the manufacturer, who agreed it's nasty stuff. But . . ."

"But what?"

"The properties in it break down and go inert twenty-four hours after the Vicane is introduced. The tent has been up and charged for about thirty-two hours now. That's close, but the manufacturer thinks none of you will be turning purple or having offspring with three arms and four eyes."

Hoffman released an exaggerated sigh of relief.

"That's the good news," Barrett pointed out.

Hoffman lifted a suspicious eyebrow, made his Ben Walton face. "What's the bad news?"

"Vicane is made by Dow Chemical," the lieutenant said.

"So?"

"They also made Agent Orange for Vietnam."

50

ROD HARRIS UNDERSTOOD SOMETHING HAD HAPPENED THE afternoon Vance Irik returned from his latest hotel arson. The investigator's face was set, grim and preoccupied, almost stunned. He had a look on his face as if he'd been burned alive.

"Vance finally nabbed his hotel pyro," John Creel announced.

The word was out. Harris watched Irik until he walked out of sight. Soon, the feeling spread that something was missing at Firehouse 2 this Christmas. It was several years since little Manny had hung around shooting hoops, but still the kid had been almost family. Christmas at the fire station meant little Manny, whether he was actually there or not.

The firefighters soon learned that Miami Beach police detectives had discovered through voice comparison and analysis that the same man had dialed 911 to report the last four hotel arson fires. As his luck would have it, Irik was busy on still another unrelated arson when police arrested the hotel pyro. After all the times he had rolled on burning hotels, he had to miss the grand finale.

A police detective paged him. "Vance, maybe you should come on down to the police station. We may have our fire starter."

Criminals who successfully eluded police while they scored crimes were sometimes attributed with supernatural powers. So often, however, the apprehension of a "super criminal" proved anticlimactic. Once caught, the perp almost always assumed his rightful proportion as some seedy little character whose luck simply ran out.

That was the way it was with the Miami Beach hotel arsonist. An alert cop noticed him hanging around the fire, had noticed him before at other fires. There would always be fire buffs, frustrated would-be firefighters, who followed the sirens. Only this guy just *looked* different. The cop followed his instincts and investigated.

"He had the odor of smoke and gasoline all in his clothing," said the police detective as he led Irik to the interrogation room. "He clammed up and won't say shit. He did say, though, that if we called you, you'd vouch for him."

Irik's eyebrows lifted.

"If you vouch for him, Vance, I guess we'll let him go. Otherwise, we're going to take another shot at cracking him."

They entered a small room, bare except for several chairs and a panel covering the back side of a one-way mirror. The detective opened the panel. Irik stared through the mirror into a second room where a thin gerbil of a man with sharp features and large brown eyes slouched forlorn in a straight-backed chair.

Irik saw more than that, however. He saw Christmas and a skinny kid, small for his age, who wore the face of a grown man laboring in coal mines, who came to shoot hoops with the firefighters, who was always present for Christmas, this semi-orphan, because as John Creel put it, this was the neediest kid any of them knew.

"But the firemen come," Manny had said that time. "The firemen always come."

Yes, they always came.

And no one else ever had. Not for Manny.

Irik turned away from the mirror. "Maybe you'd better take another shot at him," he said.

It would be weeks later, after Christmas, before Vance Irik stood in the same courtroom with Firehouse 2's Manny. Arson wasn't Manny's only conviction. The judge lumped arson in with several burglary convictions and blanketed them with a single sentence —five to ten in the state penitentiary at Raiford.

The bailiff escorted the young man from the courtroom. Irik looked at the familiar face that had changed so subtly but still so much.

Manny paused on his way out. "Firemen were my family," he said to Irik. And then he said, "Good-bye, Vance."

When he smiled wistfully, he looked, for a moment, the same as when he was ten years old.

51

Two more days until Christmas. Captain Luis Garcia stood at his office window looking into the gathering night and into the holiday lights on Pine Tree Drive that drove back the night. He sighed. He turned away from the window, but then turned back to it.

Otto Ramirez and Jim Barrett walked shoulder to shoulder across the basketball court toward the station's back door. Ramirez's blue jumpsuit was stained with blood. He looked tired. He stopped and leaned against the basketball goal post, glanced back at the sunset. Barrett continued alone into the station.

Garcia hefted a pile of reports from his desk. They told the story of firefighters like Ramirez and Barrett, of Hoffman and Mogen, Daugherty, Creel, Irik, Harris, Delfaverro, Merker, Chapman, Davidson, Guise, and all the others, and of the old-timers like Gene Spear and Jim Reilly. It would be difficult to count the number of people whose lives these firefighters had either saved or dramatically touched during Christmas holidays. People would wake up on December 25 who might not have seen that day except for firefighters.

From an age when *me* came first had emerged men and women who selflessly risked their lives in the service of others. Firefighters significantly contributed to the quality of American life—and they demanded so little in return.

They were truly, Garcia thought, America's last heroes.

The captain slipped from his office into the ready room. Mogen's elf hat had replaced the Hog's Head Chief's helmet; the Santa whiskers hung askew from the mounted animal's snout. Garcia smiled to himself as he heard Hoffman cooking in the kitchen.

"Keep nuking that crap," he heard Barrett growl, "and we'll all light up and start glowing by Christmas."

Merker joined in, wisecracking, "Update! Update! Firefighter hero and all-around good guy James Barrett has selected the perfect gift for his lovely wife. He took pity on her and is giving her a new husband."

Overcome with gentle melancholy, Garcia turned silently away and returned to his office. For him, Christmas at Firehouse 2 meant ghosts. He used the holiday season like a yardstick to measure his time with the fire department. There were so many years gone by, so fast, and so many people gone by, so fast.

In some ways, Luis Garcia's early days on the department had been difficult ones. The nation's fire departments and the solid good 'ol boy networks that ran the departments along ethnic traditions—mostly Irish or other white European stock—had been the last to crack to integration. Most fire departments still hired few blacks and even fewer women. Garcia was one of the first Latinos hired on Miami Beach. Irish Jim Reilly was his mentor.

"Fuck 'em," Reilly had said. "They have to accept you. Do your job. They'll come around."

And they had.

Jim Reilly was one of Garcia's ghosts. He died of cancer in the prime of his life.

Donald Kramer was another firehouse ghost.

Kramer and Garcia were neighborhood running mates, local kids who grew up together to make good. Garcia went from Miami Beach immigrant kid to become captain on one of the nation's finest fire departments. Not bad for a kid whose second language was English.

Kramer, on the other hand, joined the Miami Beach Police Department. Two years ago Rescue 2 received a call that began, *"Police officer shot . . ."* Garcia had had a bad feeling, so he rolled on the call with Rescue.

Some punk, a doper, had plugged Officer Kramer in the back of the head with a Saturday night special. Garcia stood a long time in the dingy tenement hallway staring through the doorway at the policeman's body floating in gore. He stood there and he couldn't move.

Paramedics kept Donald Kramer alive to the hospital; doctors kept him alive for three more days. Garcia remained at the hospital within sight of his wounded friend's bed for the entire time. Kramer had often stopped at Station 2 to visit. The day before the shooting, the two friends chatted in Garcia's office for a long time. Laughing. Talking. Reliving old times.

On the evening of Kramer's third day in the hospital, a doctor entered the room and unplugged the machines that had kept the policeman alive.

"I'm sorry," the doctor said. "He's gone."

And then there was Bob, still another ghost in Captain Garcia's firehouse. Bob was the quintessential redneck. He was not hateful in it. His manner was merely genteel idiosyncrasy, his way of quickly characterizing the changing fire department. Instead of hiding his well-bred southern prejudices, he wore them on his sleeves as badges of his own heritage. He tabbed ethnic differences, then quickly bridged any gaps caused by them by extending a big, rough, but cordial hand.

"Mexican?" he asked Garcia when they were introduced.

"Cuban."

"Uh," Bob grunted, noncommittal as he shook hands. "Never could tolerate Mexicans," he said.

Which wasn't altogether true. Bob liked everyone who took the time to penetrate that gruff front of his. Firehouse 2 returned the affection.

One afternoon there was a commotion out back where the crews washed the apparatus. Someone yelled, "Bob's sick."

By the time Garcia reached his friend's side, Bob lay on his back in the arms of another firefighter. The big man's eyes radiated pain and fear. A young firefighter called Hippie for the longish hair he wore helped a black Cuban Rescue man load the ill fireman into the back of the Rescue van. Garcia took the wheel while Hippie and the Rescue man performed CPR and injected lidocaine directly into Bob's faltering heart.

Halfway to the hospital, siren blowing, Garcia overheard swearing from the back of the van.

"No way!" he heard. "Goddamnit, man. It just can't be."

One minute Bob was helping wash engines; the next he was dead.

Captain Garcia could almost hear Bob's voice coming back over the years: "It's bad enough I had to die. What's worse is that the three guys I see last rushing me to the hospital were a spic, a nigger, and a long-haired hippie. This department has gone to shit."

But then Bob would have grinned slightly—and everyone would have forgiven him.

Ghosts.

Garcia stood looking out the window onto colorful Pine Tree. During the Christmas season, fallen firefighters assumed a presence at Firehouse 2 as real as Garcia's own presence. Garcia served as their host, their proud designate.

The sound of the alarm shook the captain from his holiday reveries. The dispatch's voice on the intercom echoed throughout the station: *"Engine Two, Engine Two, house fire at . . .*

"Rescue Two, Rescue Two, support for Engine Two . . ."

The station doors went up. Engine company roared out first, followed by the red van. Tires squalled. Sirens caught the evening breezes off the Atlantic. Whatever story Otto Ramirez had been about to tell Jim Barrett would likely remain forever untold, replaced by another story about to happen.

Upstairs, standing at the darkened window in his office, Captain Luis Garcia watched the apparatus speed south on Pine Tree. Flashing emergency lights blipped against his window. They illuminated the sudden soft smile that touched the captain's swarthy face. It was a smile of pride—pride in himself, in the firefighters of Station 2 and his department, and in the men and women everywhere who threw fate to the sirens to combat man's oldest enemy.

Firefighters made a difference. They were the stuff of heroes, of the way men were in a simpler and more heroic age, before lawyers and drugs and drive-by shootings.

They might well be, indeed, the last of the American heroes.

AWOL. Pass the buck. Blitz. Panic Button.

One of the special traits of American English is its abliity to absorb new words and metamorphize new meanings for existing words. A primary impetus for this has always been war. In his new book, *WAR SLANG*, noted word historian and author **Paul Dickson** offers the first comprehensive collection of fighting words and phrases used by Americans at war.

WAR SLANG

American Fighting Words and Phrases from the Civil War to the Gulf War

by

Paul Dickson

Available in Hardcover from

POCKET BOOKS

www.ingramcontent.com/pod-product-compliance
Lightning Source LLC
Chambersburg PA
CBHW011420070526
44584CB00026BA/3776